Using Math to Defeat the Enemy:
Combat Modeling for Simulation

by

Jeffrey S. Strickland

Simulation Educators

Colorado Springs, CO

Using Math to Defeat the Enemy: Combat Modeling for Simulation

Copyright 2011 by Jeffrey S. Strickland. All rights Reserved

ISBN 978-1-257-83225-5

Published by Lulu, Inc. All Rights Reserved

ExtendSim is a registered trademark of Imagine That, Inc.

MATLAB® and *Simulink*™ are registered trademarks of MathWorks

Wikipedia® is a registered trademark of the Wikimedia Foundation, Inc., a non-profit organization.

All pictures, unless otherwise cited, are taken from the Wikimedia Commons of the Wikimedia Foundation, and are either public domain or used under the terms of the GNU Free Documentation License or the Creative Commons Attribution/Share-Alike License. Public domain pictures have been place in public domain by the authors or their copyrights have expired.

Acknowledgement

This book is dedicated to Mariah Strickland and Evie Strickland for all their loving support.

A special thanks is extended to Laurie Parham Strickland—loving wife, nurturing mother, and loyal friend.

Foreword

This book was written for the 2nd Annual Modeling and Simulation Summit, Pre-Conference Focus Day Workshop: Using Math to Defeat the Enemy: Combat Modeling for Simulation (Monday, August 29th 2011).

Although mathematics is key to combat model for simulation, this book provides just a cursory survey of mathematical model, and includes heuristics and other algorithms. The book covers behavior models, environmental models, communication models, target detection models, attrition model, entity and aggregate level models, and multi-resolution models. It provides examples of past and current model, simulations, and wargames to stimulate learning and discussion.

Table of Contents

ACKNOWLEDGEMENT I

FOREWORD II

TABLE OF CONTENTS III

PREFACE XI

PART I. SIMULATION 1

CHAPTER 1. INTRODUCTION 3

    LEARNING OBJECTIVES 3
    KEY DEFINITIONS 4
    MODELS 4
    WARGAMES 5
        The War Aspect 6
        The Game Aspect 6
        The Simulation Aspect 6
    COMPUTER SIMULATION 7
    SIMULATION VERSUS MODELING 8
    DATA PREPARATION 9
        Input sources also vary widely: 9
        Lastly, the time at which data is available varies: 9
    TYPES 10
    APPLICATIONS OF SIMULATION 13
    INAPPROPRIATE MODELS USAGE DURING DESERT STORM 18

CHAPTER 2. SEMI-AUTOMATED FORCES 19

    US ARMY SAF EVOLUTION 19
    SIMULATOR NETWORK (SIMNET) 20
        Origination and Purpose 21
        Companies who Developed SIMNET 21
        Network Advances 22
        Graphics Advances 22
        Army Use of SIMNET for Training 23
        SIMNET Follow-On Programs 23
    MODULAR SEMI-AUTOMATED FORCES 24
        Description 25
        History 25
        Capabilities for the planning process 26

ModSAF Use 27
ONESAF 27
  Description 27
  Battlespace participants 29
  The OneSAF Agent Architecture 30
  Components 30
  Behaviors 31
  State transitions 32
  Content carriers 33
  The Blackboard 33
  Version 3.0 34
  Version 4.0 34

CHAPTER 3. SIMULATION PROTOCOLS/ARCHITECTURES 35

SIMNET 36
DISTRIBUTED INTERACTIVE SIMULATION (DIS) 36
  History 37
  Protocol data units 38
  Table 3-1. Basic DIS Concepts 40
DISTRIBUTED WARGAMING SYSTEM 41
JOINT TRAINING CONFEDERATION 43
AGGREGATE LEVEL SIMULATION PROTOCOL (ASLP) 43
  History 44
  Contributions 45
  Motivation 45
  Basic Tenets 46
  Conceptual Framework 47
  ALSP Infrastructure Software (AIS) 48
  ALSP Common Module (ACM) 48
  Time management 49
  Object management 50
  ALSP Broadcast Emulator (ABE) 51
HLA OVERVIEW 51
  Technical overview 52
  Common HLA terminology 52
  Interface specification 53
  Object model template 53
  HLA rules 54
  Base Object Model 55
  Federation Development and Execution Process (FEDEP) 55

    Distributed Simulation Engineering and Execution Process (DSEEP) ... 55
    Standards 55
    STANAG 4603 56
    DLC API 56
  COMPUTER WARGAME EVOLUTION 57
  VERIFICATION, VALIDATION AND ACCREDITATION 60
  NETCENTRIC WARFARE 61
  INFORMATION AGE WARFARE 63

PART II. MODELING 66

CHAPTER 4. ENVIRONMENTAL MODELING 67

  LEVEL OF DETAIL 68
  DATA PROCESSING 70
  STORING ENVIRONMENTAL DATA 73
  TERRAIN DATABASE TYPES 74
    Elevation By Nearest Post 75
    Elevation by interpolation for grids 76
    Interpolation with TIN 77
    Comparison of Algorithms 78
  STATIC ENVIRONMENT 78
  DYNAMIC ENVIRONMENT 81
  CLASSIC PROBLEMS IN INTERPRETATION 82
  ENVIRONMENTAL STANDARDIZATION 83

CHAPTER 5. PHYSICAL MODELING 85

CHAPTER 6. MOVEMENT MODELING 86

  GRID HOPPING 86
  SECTOR (PISTONS) MOVEMENT 87
  MOVEMENT-POINTS MOVEMENT 87
  SURFACE TILED WITH HEXAGONS 88
    Bald Earth Movement 89
  TERRAIN & FEATURE MOVEMENT 90
  PHYSICS-BASED MOVEMENT 91
  AUTOMATIC ROUTE PLANNING 91
  TOPOLOGY SMART 92
  A* SEARCH ALGORITHM 93
  GRID REGISTRATION 97
  BEYOND 2-D MOVEMENT 98
    3+3 DoF 99

6-DoF 99
    BEHAVIORAL—AGENT BASED MOVEMENT 99

CHAPTER 7. DETECTION MODELING 101

    PERFECT DETECTION 101
    GRID PROBABILITY AREAS 101
    DETECTION RANGE 102
    3D DETECTION 103
    TARGET ACQUISITION 103
    JCATS DETECTION ALGORITHM 105
    SENSOR CHARACTERISTICS 106
    TARGET CHARACTERISTICS 106
    THE BASIC TARGET ACQUISITION PROCESS 107
    LEVELS OF TARGET ACQUISITION 107
    BASIC TARGET ACQUISITION 108
    DETAILED TARGET ACQUISITION 109
    PROBABILITY OF LINE-OF-SIGHT 109
        Probability of Line-of-sight 111
        PLOS Curves 112
        Joint Warfare System Use of PLOS 113
    LINE-OF-SIGHT ALGORITHMS 114
    LINE-OF-SIGHT MODELS 114
    EXPLICIT GRID LINE-OF-SIGHT MODEL 115
    EXPLICIT SURFACE LINE-OF-SIGHT MODEL 115
    IMPLICIT INTERVISIBILITY SEGMENT LENGTH 117
    INTERMITTENT LOS MODEL 118
    DATABASE SUPPORT FOR INTERMITTENT LOS MODEL 119
    THE GLIMPSE MODEL 119
        Time-Stepped Model 119
        Estimating the Numeric Value of g 121
        Event-Stepped Model 121
    INTERMITTENT GLIMPSES 122
    CONTINUOUS LOOKING MODEL 124
    DYNTACS CURVE FIT MODEL 125
    NVEOL ACQUISITION ALGORITHM 126
    USING NVEOL 127

CHAPTER 8. COMMUNICATIONS MODELING 129

    COMMUNICATIONS MODEL EFFECTS 129
    PERFECT COMMUNICATIONS 130

- DIRECT MESSAGE PASSING 130
- BROADCAST MESSAGES 131
- COMMUNICATION NETWORK MODELS 132
  - *Mathematical Structure of Networks* *132*
  - *Basic Combat Network Structure* *133*
  - *Combat Networks* *134*
  - *Dimensions and Complexity* *135*
  - *DYNAMICS* *137*
  - *Measuring Networked Effects* *138*
  - *EVOLUTION* *141*
  - *Punctuated Growth in Complex Networks* *141*
  - *Learning and Adaptation in Complex Networks* *142*
  - *Core Shifts in Complex Networks* *144*
- OTHER NETWORK MODELS 146
  - *Neural Network Design* *146*
  - *Physics-Based Communication Networks* *147*
  - *Virtual Cell Layout (VCL)* *148*

CHAPTER 9. ENGAGEMENT MODELING – ENTITY LEVEL 150

- THE ROLE OF ATTRITION IN COMBAT MODELS 151
- POINT SYSTEM 153
- MARKOV *Pk* TABLE 154
- RANDOM NUMBERS 155
- PK'S AND RANDOM NUMBERS 156
- PRECISION ENGAGEMENTS 157
- LINEAR TARGET *Phit* 157
- SINGLE-SHOT ACCURACY 160
- RECTANGULAR TARGET *Phit* 160
- CIRCULAR TARGET *Phit* 161
- KILL CATEGORIES 162
- DIRECT-FIRE ACCURACY EXAMPLE 163

CHAPTER 10. ENGAGEMENT MODELING – AGGREGATE LEVEL 165

- LANCHESTER EQUATIONS 166
- DIFFERENTIAL EQUATIONS 167
  - *Solutions to the Lanchester Equations* *167*
  - *Steps:* *167*
  - *Initial Value Problem* *168*
- SYSTEMS OF DIFFEQ'S 168
  - *Linear Systems* *168*

 Eigen-Solutions *169*
 AGGREGATED COMBAT GROUPS 172
 FORCE RATIO ATTRITION MODELS 173
  Force Ratio Approach - Firepower Scores *174*
  Force Ratio Approach - Limitations of Firepower Scores *174*
  Force Ratio Approach - Determining Firepower Scores *175*
  Alternate Approaches *175*
  Force Ratio Approaches - Correlation of Forces *176*
  Force Ratio Attrition Models - U.S. Army Capability Analysis *177*
  Force Ratio Attrition Models - U.S. Army Capability Analysis Example *178*
  Force Ratio Attrition Models - U.S. Army Capability Analysis Example Solution *179*
  ATLAS Model *180*
  Attrition Coefficient Calculation - Potential / Anti-potential ("Eigenvalue") Method *180*
  Attrition Coefficient Calculation - Potential / Anti-potential Method. *182*
 OTHER AGGREGATED MODELS 188

CHAPTER 11. BEHAVIORAL MODELING 189

 BEHAVIORAL REPRESENTATION 189
 ARTIFICIAL INTELLIGENCE 190
 BEHAVIOR CATEGORIES 190
 AGENT TYPES 191
 BASIC INTELLIGENT AGENT 192
 AGENT CATEGORIES 193
 REACTIVE AGENT ARCHITECTURE 194
 DELIBERATE AGENT ARCHITECTURE 195
 ANATOMY OF A COGNITIVE AGENT 196
 AGENT REASONING MECHANISMS 196
 FINITE STATE MACHINES 197
 MATHEMATICAL MODELS 199
 MARKOV CHAINS 199
 FUZZY LOGIC 200
 FUZZY RULES 201
  Neural Networks *202*
 NEURAL NETWORKS IN FIELDS OF BATTLE 203
 GENETIC ALGORITHMS (GA) 204
  The Algorithms *205*
  Genetic Algorithms in bSerene *206*
  Genetic Algorithms in Return Fire *207*

Path Planning Mechanism using Genetic Algorithms 208
EVOLVING NEURAL NETWORKS (ENN) 209
   ENN in MANA *209*
EXPERT SYSTEMS 211
   CLIPS & Derivatives *212*
MILITARY ENTERTAINMENT COMPLEX 214
   America's Army (early versions) *215*
   Unreal Engine 2 (America's Army v1.0 ~ v2.x) *216*
   America's Army 3 .... *216*
   Government applications *217*
   Full Spectrum Warrior *217*
   Development *218*
SOAR ARCHITECTURE 220
   TacAir-Soar *222*
   Tank Soar *222*
ARTIFICIAL PROBLEM SOLVERS: SWARM INTELLIGENCE 223
   Swarming Characteristics *223*
   Routing in Networks *225*
   Sugarscape *226*
   EINSTein *227*
   Interactive "Tool Box" / Laboratory *229*
   MARSS *230*
   MARSS features *231*
SAMPLE PROJECTS 231

CHAPTER 12. MULTI-RESOLUTION MODELING 233

MULTI-RESOLUTION MODELING 233
COMMON RESOLUTION LEVELS 234
AGGREGATION 234
EFFECTS OF AGGREGATION 235
PROPER AGGREGATION 235
FORMAL AGGREGATION 236
   Extensive Aggregation *237*
   Intensive Aggregation *237*
HIERARCHICAL ATTRITION ALGORITHMS 238
   Hierarchical Attrition Algorithms - ATCAL *239*
   The Resulting Aggregated Attrition Parameters *240*
   High-Resolution Parameters: *240*
   Development of the Point Fire Attrition Equation *241*
WHAT IS MULTI-RESOLUTION MODELING? 242

MRM: VEHICLE LOCATION AND HEALTH 243
THE MRM PROBLEM 244
   Aggregate and Entity *245*
   MRM Air Combat Example *246*
MRM ISSUES 246
   Consistent History *246*
   Movement *247*
   Light-of-Sight *248*
   Spreading Disaggregation *249*
DIRECT FIRE SOLUTIONS 249
   Firewall: Direct Contact *250*
   Firewall: Partial Directed *250*
   Firewall: Horizon of Interest *251*
AGGREGATE LOCATION 252
MEASURE OF CONSISTENCY 253
MULTIPLE LEVELS OF RESOLUTION 254
DISAGGREGATION LAYERS 255
BATTLEFIELD METAPHYSICS 257
HOMELAND SECURITY & HUMANITARIAN OPERATION 258
COMMERCIAL-OFF-THE-SHELF (COTS) SIMULATIONS 258
   TacOps 4 .... *258*
   Brigade Combat Team *259*
   Decisive Action *260*
REFERENCES 263

WORKS CITED **263**
INDEX **273**

Preface

Many of the criticisms directed towards military simulations derive from an incorrect application of them as a predictive and analytical tool. The outcome supplied by a model relies to a greater or lesser extent on human interpretation and therefore should not be regarded as providing a 'gospel' truth. However, most game theorists and analysts generally understand this, it can be tempting for a layman—for example, a politician who needs to present a 'black and white' situation to his electorate—to settle on an interpretation that supports his preconceived position. Tom Clancy, in his novel *Red Storm Rising*, illustrated this problem when one of his characters, attempting to persuade the Soviet Politburo that the political risks of war with NATO were acceptable, used as evidence the results of a simulation carried out to model just such an event. It is revealed in the text that there were in fact three sets of results from the simulation; a best-, intermediate- and worst-case outcome. The advocate of war chose to present only the best-case outcome, thus distorting the results to support his case (Clancy, 1988).

There have been many charges over the years of computerized models being unrealistic and slanted towards a particular outcome. Critics point to the case of military contractors, seeking to sell a weapons system. For obvious reasons of cost, weapons systems (such as an air-to-air missile system for use by fighter aircraft) are modeled extensively on computers. Without testing of their own, a potential buyer must rely to a large extent on the manufacturer's own model. This might well indicate a very effective system, with a high kill probability (P_k). However, it may be the model was configured to show the weapons system under ideal conditions, and its actual operational effectiveness will be somewhat less than stated. The US Air Force quoted their AIM-9 Sidewinder missile as having a P_k of 0.98 (it will successfully destroy 98% of targets it is fired at). In operational use during the Falklands War in 1982, the British recorded its actual P_k as 0.78 (Allen T. B., 1987).

Human factors have been a constant thorn in the side of the designers of military simulations. Whereas political-military simulations are often required by their nature to grapple with what modelers refer to as "soft" problems, purely military models often seem to prefer to concentrate on hard numbers. While a warship can be regarded, from the perspective of a model, as a single entity with known parameters (speed, armor, gun power, and the like), land warfare often depends on the actions of small groups or individual soldiers where training, morale, intelligence, and personalities (leadership) come into play. For this reason, it is more taxing to model—there are many difficult-to-formulate variables. One valid criticism of some military simulations is these nebulous human factors are often ignored (partly because they are so hard to model accurately). Other perplexing issues include aggregation-disaggregation, communication networks, attrition, and end-game modeling.

"Using Math to Defeat the Enemy: Combat Modeling for Simulation" is intended to provide a foundation in the underlying combat modeling issues of military simulations. Of course, this is just a background, and a more rigorous treatment can be found in my book, Mathematical Modeling of Warfare and Combat Phenomenon (2011), Lulu.com, ISBN 978-1-4583-9255-8. Ultimately, this is a resource/reference book covering a wide gambit of military modeling issues.

This book is organized in two parts: Simulation (Part I) and Modeling (Part II). There are numerous practical applications and example models used in past and current military simulations.

This book is a result of about 25 years of use, application, research, and teaching military modeling and simulation. Much of the material in this book based on practical experience with modeling and simulation and extraction of my course notes from PowerPoint presentations.

Jeffrey S. Strickland, Ph.D.
CMSP, ASEP
President
Simulation Educators
Colorado Springs, Co

www.simulation-educators.com

Part I. Simulation

Simulation is the imitation of some real thing, state of affairs, or process. The act of simulating something generally entails representing certain key characteristics or behaviors of a selected physical or abstract system. Simulation is used in many contexts, such as simulation of technology for performance optimization, safety engineering, testing, training, education, and video games. Training simulators include flight simulators for training aircraft pilots in order to provide them with a lifelike experience. Simulation is also used for scientific modeling of natural systems or human systems in order to gain insight into their functioning (Smith R. , Simulation Article, 1998). Simulation can be used to show the eventual real effects of alternative conditions and courses of action. Simulation is also used when the real system cannot be engaged, because it may not be accessible, or it may be dangerous or unacceptable to engage, or it is being designed but not yet built, or it may simply not exist (Sokolowski & Banks, 2009)

The term Military Simulation can be used to cover a wide spectrum of activities, ranging from full scale field exercises, to abstract computerized models that can proceed with little or no human involvement such as the Rand Strategy Assessment Center (RSAC) (Hall, Shapiro, & Shukia, 1993).

Full-scale military exercises, or even smaller scale ones, are not always feasible or even desirable. Availability of resources, including money, is a significant factor—it is an expensive endeavor to release men and materiel from any standing commitments, transport them to a suitable location, and then cover additional expenses such as Petroleum Oil Lubricants (POL) usage, equipment maintenance, supplies and consumables replenishment and other items. In addition, certain warfare models are not amenable to verification using this realistic method. It would, for example, be impossible to accurately test an attrition scenario by killing one's own troops.

The forerunner of modern simulations was the Prussian game Kriegspiel, which appeared around 1811 and is sometimes credited with the Prussian victory in the Franco-Prussian War (Caffrey, 2000). It was distributed to each Prussian regiment and they were ordered to play it regularly, prompting a visiting German officer to declare in 1824, "It's not a game at all! It's training for war!" (Wilson A. , 1968) Eventually so many rules sprang up, as each regiment improvised their own variations, two versions came into use. One, known as "rigid Kriegspiel", was played by strict adherence to the lengthy rulebook. The other, "free Kriegspiel", was governed by the decisions of human umpires (Matute, 1970). Each version had its advantages and disadvantages: rigid Kriegspiel contained rules covering most situations, and the rules were derived from historical battles where those same situations had occurred, making the simulation verifiable and rooted in observable data, which some later American models discarded. However, its prescriptive nature acted against any impulse of the participants towards free and creative thinking. Conversely, free Kriegspiel could encourage this type of thinking, as its rules were open to interpretation by umpires and could be adapted during operation. This very interpretation, though, tended to negate the verifiable nature of the simulation, as different umpires might well adjudge the same situation in different ways, especially where there was a lack of historical precedent. In addition, it allowed umpires to weight the outcome, consciously or otherwise.

The above arguments are still cogent in the modern, computer-heavy military simulation environment. There remains a recognized place for umpires as arbiters of a simulation, hence the persistence of manual simulations in war colleges throughout the world. Both computer-assisted and entirely computerized simulations are common as well, with each being used as required by circumstances.

Ideally, military simulations should be as realistic as possible—that is, designed in such a way as to provide measurable, repeatable results that can be confirmed by observation of real-world events. This is especially true for simulations that are stochastic in nature, as they are used in a manner that is intended to produce useful, predictive outcomes. Any user of simulations must always bear in mind that they are, however, only an approximation of reality, and hence only as accurate as the model itself.

Key issues in simulation include acquisition of valid source information about the relevant selection of key characteristics and behaviors, the use of simplifying approximations and assumptions within the simulation, and fidelity and validity of the simulation outcomes.

Chapter 1. Introduction

Mathematics plays a critical role in many modeling and simulation applications, and whenever detailed representations of combat phenomenon are required (e.g., attrition, detection, etc.). The required mathematics can seem daunting, but a wealth of problems relevant to the modeling and simulation (M&S) community can be addressed with efficient numerical algorithms and heuristics. We shall present a small sampling of these methods and relevant examples of their usage. Examples include combat attrition, target engagement, detection, and missile defense. Some of these can be implemented with a spreadsheet or Commercial-off-the-Shelf (COTS) simulation product.

Learning Objectives

1. Describe the scope of mathematical and heuristic combat models.
2. Compare and contrast different representations of combat phenomenon.
3. List combat behaviors that can be represented by mathematical & heuristic models.
4. State the various types of mathematical and heuristic combat models.
5. Identify examples of mathematical and heuristic combat models.

Combat Simulations and Simulators play a major role in today's military in training, exercises, and military operations (TEMO); test & evaluation; research, development and, acquisition (RDA); requirements, and concepts (ACR). The military has a number of standing models, simulations, and simulators that serve a variety of purposes. An understanding of the modeling techniques and technologies used to build these tools is essential.

Key Definitions

A **model** is a physical, mathematical, or otherwise logical representation of real-world system, entity, phenomenon, or process. A model is something you get by going out into the real-world and grabbing information about a real world system or process and dragging it back into the simulation and writing it in software.

A **simulation** is the imitation of the operation of a real-world process or a system over time. It is the interaction of models as they are operating. The execution of the software that you have written is a simulation.

A **wargame** is a simulation of competitive operations, involving two or more opposing teams, using rules, data, and procedures designed to depict an actual or assumed real-life situation. A wargame is an event that is bigger than the hardware and software. It involves the people, the equipment, and the coordination it takes to teach people to live in the virtual world that has been created for them.

Models

A **model** is a logical description of how a system performs. Simulations involve designing a model of a system and carrying out experiments on it as it progresses through time. For example, the board game *Monopoly* is a model of a real system– the hotels and facilities of Atlantic City. When you play *Monopoly*, you are simulating that system. Simulation with computers means that instead of interacting with a real system, you create a model, which corresponds to it in certain aspects.

You can use a model to describe how a real-world activity will perform. Models also enable you to test hypotheses at a fraction of the cost of actually undertaking the activities which the models simulate. For example, if you are a hardware designer, you can use computers to simulate the performance of a proposed system before building it.

Table 1-1. Model means different things to different people

• Mathematic equations describing the behavior of a system.	$\dfrac{dx}{dt} = ay, \dfrac{dy}{dt} = bx$
• Algorithms of the physical behavior of an object.	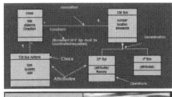
• The software that captures the algorithms.	
• 3D digital image of visual object.	
• Strictly speaking, a model refers to the item that is created to mimic the real world.	

One of the principal benefits of a model is that you can begin with a simple approximation of a process and gradually refine the model as your understanding of the process improves. This "step-wise refinement" enables you to achieve good approximations of very complex problems surprisingly quickly. As you add refinements, your model becomes more and more accurate.

Wargames

Depending on one's definition, "**wargaming**" might extend from the spectacle that took place in the Coliseum to the abstract exercises of chess. Wargames share three properties, which are present in various levels, depending on the game and genre of games in question. Wargames date back to 1664.

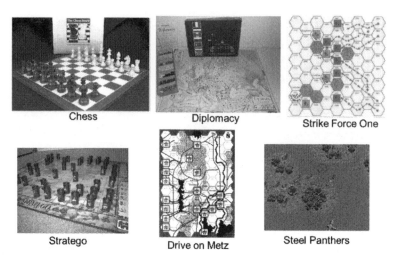

Figure 1-1. Examples of wargames

The War Aspect

Obviously, not all games are related to war. Wargames are a subcategory of games that simulate the activity of war and borrow war's vocabulary. Chess might be characterized as a borderline wargame. Chess borrows something from war, or at least war as it was practiced at one point in time. Political-military simulations and board games like *Diplomacy* are also somewhat borderline as wargames. Undoubtedly, they involve military activities, but sometimes military issues are not a predominant focus of the games.

The Game Aspect

Not all war-related activities short of war can be described as wargames. For example, although military exercises (maneuvers) are often described as wargames, there is often very little game in the activity. Gaming inevitably calls upon participants to compete against themselves or others in pursuit of some defined goal. Usually this forces players to think and act in new and creative ways, anticipating future events and devising strategies to handle them. Simulations of war, which do not provide such an opportunity for strategic thinking, are not wargames.

The Simulation Aspect

Good wargames often prepare and educate participants for the activity of war in some sense through exposing them to a simulation of some aspect of actual war. Though the value (and existence) of this last dimension is subject to debate, the educational aspect of a wargame would seem to depend on the validity of the game's model. Good games attempt to simulate the conditions and realities of their subject matter. If these attempts are successful, then playing wargames should (at least to some extent) help one prepare for and understand warfare, without the cost of actual deaths. The better the accuracy of the simulation, the better the education that might be gained. Indeed, the United States military (and other military bodies throughout history) have historically supported the value of wargames as a learning and training tool for this reason.

Computer simulation

A computer simulation, a computer model, or a computational model is a computer program, or network of computers, that attempts to simulate an abstract model of a particular system. Computer simulations have become a useful part of mathematical modeling of many natural systems in physics (computational physics), astrophysics, chemistry and biology, human systems in economics, psychology, social science, and engineering. Simulations can be used to explore and gain new insights into new technology, and to estimate the performance of systems too complex for analytical solutions (Strogatz, 2007).

Computer simulations vary from computer programs that run a few minutes, to network-based groups of computers running for hours, to ongoing simulations that run for days. The scale of events being simulated by computer simulations has far exceeded anything possible (or perhaps even imaginable) using the traditional paper-and-pencil mathematical modeling. Over 20 years ago, the DoD High Performance Computer Modernization Program simulated a desert-battle of one force invading another. This involved the modeling of 66,239 tanks, trucks and other vehicles, on simulated terrain around Kuwait, using multiple supercomputer (Watson, 1997). Additionally, Los Alamos National Laboratories built a 1-billion-atom model of material deformation in 2002, and a 2.64-million-atom model of the complex maker of protein in all organisms, a ribosome, in 2005 (Ambrosiano, 2005). The Blue Brain project at École Polytechnique in Lausanne (Switzerland), began in May 2005, to create the first computer simulation of the entire human brain, right down to the molecular level (Graham-Rowe, 2005).

Simulation versus modeling

Traditionally, forming large models of systems has been via a mathematical model, which attempts to find analytical solutions to problems and thereby enable the prediction of the behavior of the system from a set of parameters and initial conditions.

While computer simulations might use some algorithms from purely mathematical models, computers can combine simulations with reality or actual events, such as generating input responses, to simulate test subjects who are no longer present. Whereas the missing test subjects are being modeled/simulated, the system they use could be the actual equipment, revealing performance limits or defects in long-term use by these simulated users.

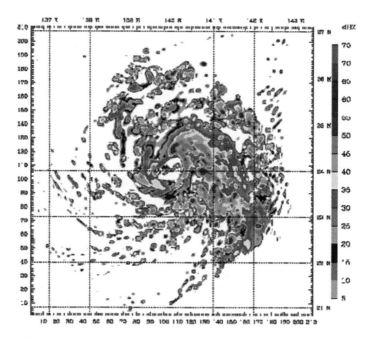

Figure 1-2. A 48-hour computer simulation (one frame) of Typhoon Mawar using the Weather Research and Forecasting model

Note that the term computer simulation is broader than computer modeling, which implies that all aspects are being modeled in the computer representation. However, computer simulation also includes generating inputs from simulated users to run actual computer software or equipment, with only part of the system being modeled: an example would be flight simulators that can run machines as well as actual flight software.

Computer simulations are used in many fields, including science, technology, entertainment, health care, and business planning and scheduling.

Data preparation

The external data requirements of simulations and models vary widely. For some, the input might be just a few numbers (for example, simulation of a waveform of AC electricity on a wire), while others might require terabytes of information (such as weather and climate models).

Input sources also vary widely:

- Sensors and other physical devices connected to the model;
- Control surfaces used to direct the progress of the simulation in some way;
- Current or Historical data entered by hand;
- Values extracted as by-product from other processes;
- Values output for the purpose by other simulations, models, or processes.

Lastly, the time at which data is available varies:

- "invariant" data is often built into the model code, either because the value is truly invariant (e.g. the value of π) or because the designers consider the value to be invariant for all cases of interest;
- data can entered into the simulation when it starts up, for example by reading one or more files, or by reading data from a preprocessor;
- data can be provided during the simulation run, for example by a sensor network;

Because of this variety, and that many common elements exist between diverse simulation systems, there are a large number of specialized simulation languages. The best-known of these must be Simula (sometimes Simula-67, after the year 1967 when it was proposed). There are now many others.

Systems accepting data from external sources must be very careful in knowing what they are receiving. While it is easy for computers to read in values from text or binary files, what is much harder is knowing what the accuracy (compared to measurement resolution and precision) of the values is. Often it is expressed as "error bars", a minimum and maximum deviation from the value seen within which the true value (is expected to) lie. Because digital computer mathematics is not perfect, rounding and truncation errors will multiply this error up, and it is therefore useful to perform an "error analysis" (Taylor J. R., 1999) to check that values output by the simulation are still usefully accurate.

Even small errors in the original data can accumulate into substantial error later in the simulation. While all computer analysis is subject to the "GIGO" (garbage in, garbage out) restriction, this is especially true of digital simulation. Indeed, it was the observation of this inherent, cumulative error, for digital systems that is the origin of chaos theory.

Types

Computer models can be classified according to several independent pairs of attributes, including:

- **Stochastic or deterministic** (and as a special case of deterministic, chaotic)
- **Steady-state or dynamic**
- **Continuous or discrete** (and as an important special case of discrete, discrete event or DE models)
- **Local or distributed**.

Equations define the relationships between elements of the modeled system and attempt to find a state in which the system is in equilibrium. Such models are often used in simulating physical systems, as a simpler modeling case before dynamic simulation is attempted.

Dynamic simulations model changes in a system in response to (usually changing) input signals.

Stochastic models use random number generators to model chance or random events.

Discrete Event Simulation (DES) manages events in time. Most computer, logic-test and fault-tree simulations are of this type. In this type of simulation, the simulator maintains a queue of events sorted by the simulated time they should occur. The simulator reads the queue and triggers new events as each event is processed. It is not important to execute the simulation in real time. It is often more important to be able to access the data produced by the simulation, to discover logic defects in the design, or the sequence of events.

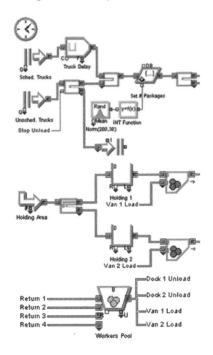

This figure depicts portions of a transportation model implement as a discrete event simulation using *ExtendSim* 8. Management wanted a model of this system because it believed that packages were taking too long to be processed at the terminal and the terminal in general was not operating efficiently.

An efficiently operating terminal would: (1) Have trucks to immediately start unloading at a dock when they arrive. (2) Delivery vans would also not have to wait to load their packages and begin delivering them. (3) Workers would be busy most of the time, and the system would be as inexpensive as possible to operate (Strickland, Discrete Event Simulation Using ExtendSim 8, 2010).

Figure 1-3. Example of a discrete event simulation

A **continuous dynamic simulation** performs numerical solution of differential-algebraic equations or differential equations (either partial or ordinary). Periodically, the simulation program solves all the equations, and uses the numbers to change the state and output of the simulation. Applications include flight simulators, construction and management simulation games, chemical process modeling, and simulations of electrical circuits. Originally, these kinds of simulations were actually implemented on analog computers, where the differential equations could be represented directly by various electrical components such as op-amps.

By the late 1980s, however, most "analog" simulations were run on conventional digital computers that emulate the behavior of an analog computer.

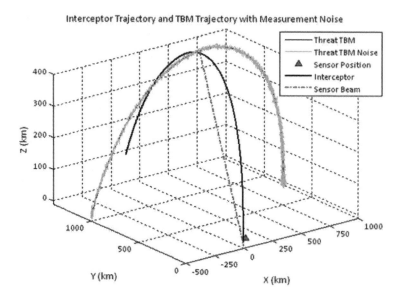

Figure 1-4. This figure shows the plots produced by a continuous dynamic simulation in MATLAB® and Simulink™. A single surface-to air missile defended against a single threat theater ballistic missile (TBM). The TBM plot includes signal noise. The small triangle represents the sensor location (Strickland, Missile Flight Simulation: Surface-to-Air Missiles, 2010).

Agent-based simulation is a special type of discrete simulation that does not rely on a model with an underlying equation, but can nonetheless be formally represented. In agent-based simulation, the individual entities (such as molecules, cells, trees or consumers) in the model are represented directly (rather than by their density or concentration) and possess an internal state and set of behaviors or rules that determine how the agent's state is updated from one time-step to the next.

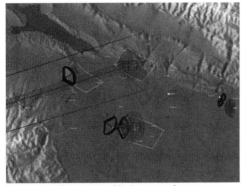

Figure 1-5. SEAS screenshot

System Evaluation and Analysis Simulation (SEAS) was used to support a Space and Missile defense Command study of space radar's contribution to a Future Combat System vignette in 2006. SEAS is a Government Off-The-Self (GOTS) agent-based simulation (Strickland, An Agent-Based Simulation of Satellite Utility For the US Army Future Force, 2006).

Distributed models run on a network of interconnected computers, possibly through the Internet. Simulations dispersed across multiple host computers like this are often referred to as "distributed simulations". There are several standards for distributed simulation, including Aggregate Level Simulation Protocol (ALSP), Distributed Interactive Simulation (DIS), the High Level Architecture (simulation) (HLA) and the Test and Training Enabling Architecture (TENA).

Applications of Simulation

- Aid to thought
- Communication
- Training
- Entertainment
- Experimentation
- Prediction

Dr. Salah E. Elmaghraby

In 1968, Dr. Salah E. Elmaghraby published a list of the applications of simulation. This still serves as an excellent summary of the general categories (Elmaghraby, 1968).

Figure 1-6. Dr. Salah E. Elmaghraby's applications of simulation

- It can be used to assist you in thinking about a problem. People who consider the operation of complex systems have found it difficult to gather all the information in their minds, hold them there, and work out the details, without the aid of some system. You can use a simulation to capture those ideas and you it as part of your brain to study the system of interest.

- It can be an aid in communicating ideas to another person. Once you capture the information required to describe the system, you can use simulation to communicate those ideas with others and to show them how the system operates.
- It is valuable for training people to perform tasks or to understand new ideas. You can use a simulation to place people in a virtual simulation that requires them to work their way through various virtual situations and learn from them.
- It is a tool for predicting the future of a system based on information about the past and present. It can show the emergent behaviors buried in the data.
- It can aid scientists in conducting experiments.
- And, adding to Dr. Elmaghraby's list, it is a wonderful form of entertainment.

ACR
- Experiments with new concepts and advanced technologies to develop requirements in doctrine, training, leader development, organizations, materiel, and soldiers
- Evaluates the impact of horizontal technology integration through simulation and experimentation

RDA
- Designs, develops, and acquires weapons systems and equipment
- Performs scientific inquiry to discover or revise facts and theories of phenomena, followed by transformation of these discoveries into physical representations

TEMO
- Includes most forms of training at echelons from the individual soldier through collective, combined arms, joint and/or combined exercises
- Includes mission rehearsals and evaluations of all phases of war plans
- Includes analysis during the rehearsal or evaluation to validate the plan

Advanced Concept Requirements (ACR). ACR experiments with new concepts and advanced technologies to develop requirements in doctrine, training, leader development, organizations, materiel, and soldiers. They evaluate the impact of horizontal technology integration through simulation and experimentation.

Research, Development, and Acquisition (RDA). RDA designs, develops, and acquires weapons systems and equipment. They perform scientific inquiry to discover or revise facts and theories of phenomena, followed by transformation of these discoveries into physical representations.

Training, Exercise, Military Operations (TEMO). These include most forms of training at echelons from the individual soldier through collective, combined arms, joint and/or combined exercises. They also include mission rehearsals and evaluations of all phases of war plans, and analysis during the rehearsal or evaluation to validate the plan.

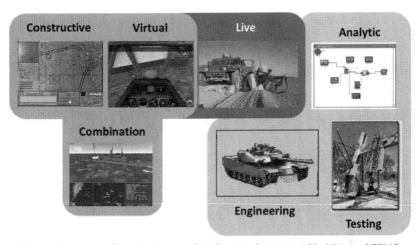

Figure 1-7. Types of simulations used in the tree domains: ACR, RDA, and TEMO

Within the Department of Defense, simulations have emerged and evolved on all fronts. As a result, it has been very difficult to compare or categorize them. However, in 1992 the Defense Science Board defined the major categories of defense simulations, greatly easing communication within the community. No simulation fits perfectly into any one category, but each simulation is usually dominated by the characteristics of one of them.

Constructive Simulation is the use of wargames to improve command and staff level decision-making. It is a mental experience of third-person perspective.

Virtual Training is the use of computer graphics to stimulate soldiers operating combat equipment. It is sensory experience of first-person perspective.

Live Simulation is the staging of mock combat experiences in a relatively non-lethal environment. It is a very physical experience of first-person perspective.

Analytical Simulation is the use of models for the purposes of understanding the performance of a system.

Engineering Simulation replicates the performance of individual pieces of equipment for the purposes of improving its performance.

Testing Simulation is the use of virtual environments to stimulate real equipment to effectively measure its performance characteristics.

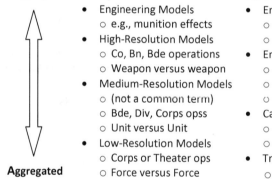

Disaggregated

Army
- Engineering Models
  - e.g., munition effects
- High-Resolution Models
  - Co, Bn, Bde operations
  - Weapon versus weapon
- Medium-Resolution Models
  - (not a common term)
  - Bde, Div, Corps opss
  - Unit versus Unit
- Low-Resolution Models
  - Corps or Theater ops
  - Force versus Force

Air Force
- Engineering Models
  - e.g., munition effects
  - Milliseconds to seconds
- Engagement-level
  - Aircraft and ADA interactions
  - Missile versus A/C
  - Minutes to hours
- Campaign-Level
  - Air Campaign
  - Multiple days of operations
- Training Comunity
  - Entity-Based/Aggregate

Aggregated

Figure 1-8. Every armed service may have different simulation taxonomies.

The US Army and USAF have different but parallel taxonomies that vary according to resolution and aggregation. These concepts will be further discussed as a part of multi-resolution modeling.

The Army Operations Research Office (ORO) at Johns Hopkins University created the first computerized simulation in 1948. A team lead by R.P. Rich and Alfred Hausrath created an anonymous "air defense simulation" which they used to study North American air defense capabilities and naval anti-aircraft guided missile systems. This simulation operated on one of the first Univac computers.

ORO also created the very first digital computer simulation--the Computerized Monte Carlo Simulation (CARMONETTE) in 1953. It represented a company or battalion sized battle in which units were able to Move, Prepare to Fire, and Fire. This simulation became operational in 1956 and was first used to study tank/anti-tank engagements. In 1960, CARMONETTE II gained the ability to represent infantry, followed by CARMONETTE III in 1966 with armed helicopter support, and CARMONETTE IV, which added communications and night vision effects.

First Computerized Wargames (Strickland, Introduction, 2005)

"Air Defense Simulation"
- Hosted on the Univac computer
- North American air defense
- Naval anti-aircraft guided missiles

CARMONETTE
- 1953 Computerized Monte Carlo Simulation
- Operational 1956-1970
- Tank/Anti-Tank (v.I), Infantry (v.II), Helicopters (v.III), Communications(v.IV)

Army Operations Research Office (ORO) at Johns Hopkins University

Remington Rand employees, Harold E. Sweeney (left) and J. Presper Eckert (center) demonstrate the U.S. Census Bureau's UNIVAC for CBS reporter Walter Cronkite (right)

UNIVAC I

Note: In 1952 the Univac received world-wide recognition by predicting the presidential victory of Dwight D. Eisenhower over Adali Stevenson. Political analysts were betting on Stevenson, but the Univac extrapolated early election returns and concluded that Eisenhower would be a strong winner.

Inappropriate models usage during Desert Storm

The very respected Concepts Evaluation Model (CEM) was a piston model which we were using to determine how would our forces would push Iraqi units out of Kuwait. It was not very accurate

Another model, the very respected Extended Air Defense Simulation (EADSIM), was used to try to calculate our effectiveness of our daily airstrikes against targets. EADSIM was not calibrated for the kind of forces we would fight during Desert Storm, and the results were being tweaked a little bit every day until that could get the model behave like reality.

Neither one of these model projects were embarrassed by this, and published article in a Military Operations Research Society (MORS) publication Warfare Modeling (Smith R. , Military Simulation Techniques & Technology, 2006).

Table 1-4. Inappropriate models usage during Desert Storm

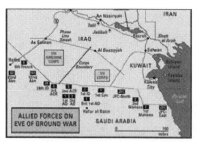

Concepts Evaluation Model (CEM)
- Army Concepts Analysis Agency
- Two-sided deterministic piston model
- Air effects as input modifications

Extended Air Defense Simulation (EADSIM)
- Air Force Studies and Analysis Agency
- Evaluate air defense networks
- Calibrated to match mission results each day

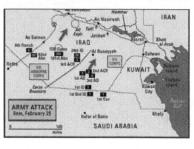

Corps Battle Simulation (CBS)
- US Central Command
- Estimate attrition in ground warfare
- Compare tactical options

All of these models assumed Soviet-type tactics and equipment and did not capture the "flavor" of Iraqi combat.

The Corps Battle Simulation (CBS) was being used to determine the level of attrition that would occur to US and Iraqi forces. It overestimated US casualties by several order of magnitude, because the assumptions behind these models could not be applied to the kind of combat experienced in Desert Storm. The assumptions behind these models were based on a soviet-bloc force.

Chapter 2. Semi-Automated Forces

A semi-automated force (SAF) is a computer-generated force (CGF) with doctrinally and tactically correct behaviors built into the software. The SAF can operate automatically (turn the simulation on and let it run a scenario to completion), or a human operator can intervene to alter the automated behaviors.

US Army SAF Evolution

During the 1980's, the Defense Advanced Research Projects Agency (DARPA) developed virtual simulators for the M1 Abrams main battle tank and the M2 Bradley infantry-fighting vehicle to provide a sophisticated crew training capability. These simulators, each the replica of a single vehicle crew compartment, were then integrated on a local area network as a Simulation Network (SIMNET) to permit collective command and control (C2) and maneuver training of small units (armor platoon, etc.). The Army established SIMNET facilities in CONUS, USAREUR, and Korea.

As SIMNET users gained experience they needed a more robust and dynamic threat. DARPA therefore developed a constructive simulation to provide that opposing force (OPFOR) for users of SIMNET. This simulation used a semi-automated force (SAF) which could be task organized and given orders to execute basic military missions (i.e. attack, defend, etc.). This SIMNET SAF was an improvement but too inflexible. The SAF had been written in "hard code" and each time a change or improvement was made to the software, the SAF required 100 percent recompilation, which required downtime and consumed man-hours and money.

DARPA therefore developed ModSAF. ModSAF means "modular semi-automated forces". ModSAF version 1.0 was an improvement over SIMNET SAF because each major area of the software code (force characteristics, terrain, operating parameters, etc.) was now a separate module. Changes could be focused on the specific area requiring improvement and recompilation time was reduced.

Due to work by individual agencies on the different versions of ModSAF fielded for testing or use, there are many different variations available. Most of these versions have been developed by individual agencies to respond to agency-unique requirements. Users make changes to a version of ModSAF, as required, and the software modules in one user's version no longer resembles similar modules in another agency's version. The baseline versions (1.0, 2.0, and 3.0) of ModSAF exist under strict configuration control of STRICOM, the ModSAF material developer.

In 1998, the OneSAF program began work on the OneSAF Testbed Baseline (OTB). The OTB effort focused on drawing the best attributes from various ModSAF versions, improving behaviors, increasing code efficiency, and experimenting with new technology. OneSAF was formalized as an acquisition program in April 2000 with its designation as an Acquisition Category (ACAT) III program and delegation of the Milestone Decision Authority (MDA) to Commanding General (CG), Army Materiel Command (AMC). The Army Acquisition Executive (AAE) formally chartered PM OneSAF to manage all OneSAF efforts on 01 May 2000. On 16 May 2000, the MDA responsibility was delegated to the Commander STRICOM (subsequently re-designated as PEO STRI). OTB version 1.0 was released as the replacement for ModSAF in January 2001. In December 2003 Milestone B/C was approved by the MDA and the program moved forward with development of the Full Operational Capability (FOC) baseline. FOC was met in March 2006 and fielding was initiated in September 2006 with the release of OneSAF Version 1.0. Additional versions continue to be fielded as Pre-Planned Product Improvements (P3I) and Co-Developer handovers are integrated into the baseline. Version 2.0 was released in February 2008, Version 3.0 in February 2009, Version 4.0 in April 2010 and Version 5.0 in September 2010. Recently, OneSAF reached the ACAT II status. It is currently in the production and support phase of the life-cycle with ongoing software production and implementation of Pre-Planned Product Improvements (P3I), as approved by the TRADOC Project Office (TPO) OneSAF.

Simulator Network (SIMNET)

SIMNET was a wide area network with vehicle simulators and displays for real-time distributed combat simulation: tanks, helicopters and airplanes in a virtual battlefield. SIMNET was developed for and used by the United States military. SIMNET development began in the mid-1980s, was fielded starting in 1987, and was used for training until successor programs came online well into the 1990s.

Figure 2-1. Screenshots from SIMNET

Origination and Purpose

Jack Thorpe of the Defense Advanced Research Projects Agency (DARPA) saw the need for networked multi-user simulation. Interactive simulation equipment was very expensive, and reproducing training facilities was likewise expensive and time consuming. In the early 1980s, DARPA decided to create a prototype research system to investigate the feasibility of creating a real-time distributed simulator for combat simulation. SIMNET, the resulting application, was to prove both the feasibility and effectiveness of such a project (Pimental & Blau, 1994).

Training using actual equipment was extremely expensive and dangerous. Being able to simulate certain combat scenarios, and to have participants remotely located rather than all in one place, hugely reduced the cost of training and the risk of personal injury (Rheingold, 1992). Long-haul networking for SIMNET was run originally across multiple 56 k-bit/s dial-up lines, using parallel processors to compress packets over the data links. This traffic contained not only the vehicle data but also compressed voice.

Companies who Developed SIMNET

Three companies developed SIMNET: Delta Graphics, Inc.; Perceptronics, Inc.; and Bolt, Beranek and Newman (BBN), Inc. There was no prime contractor on SIMNET; independent contracts were let directly to each of these three companies. BBN developed the vehicle simulation and network software, as well as other software such as artillery, resupply, and semi-automated forces often used for opposing forces. Delta Graphics, based in Bellevue, Washington, developed the graphics system and terrain databases. BBN eventually bought Delta Graphics was. Perceptronics, based in Los Angeles, was responsible for the actual SIMNET simulators.

21

The company's engineers, human factors personnel and manufacturing team designed, developed and built over 300 full-crew simulators, integrating the controls, sound systems and visual systems into the special simulator shells; they also installed the simulators in a number of facilities in the US and Germany, trained the operators and supported the system for several years.

Network Advances

Since this was a networked simulation, each simulation station needed its own display of the shared virtual environment. The display stations themselves were mock-ups of certain tank and aircraft control simulators, and they were configured to simulate actual conditions within the actual combat vehicle. The tank simulators, for example, could accommodate a full four-person crew complement to enhance the effectiveness of the training. The network was designed to support up to several hundred users at once. The fidelity of the simulation was such that it could be used to train for mission scenarios and tactical rehearsals for operations performed during the U.S. actions in Desert Storm in 1992 (Robinett, 1994).

The system used the concept of "dead reckoning" to collect the positions of the objects and actors within the simulated environment. Essentially this approach proposes that the current position of an object can be calculated from its previous position and velocity (which is composed of vector and speed elements) (Pimental & Blau, 1994). Its use in the Gulf War demonstrates the success of the SIMNET, and its legacy was viewed as proof that real-time interactive networked cooperative virtual simulation is possible for a large user population. Later, the Terrestrial Wideband Network (a high speed descendant of the ARPANET that ran at T1 speeds) was used to carry traffic. This network remained under DARPA after the rest of ARPANET was merged with NSFNet and the ARPANET was decommissioned (Rheingold, 1992).

Graphics Advances

In addition to the network, the second fundamental challenge at the time SIMNET was conceived was the inability of graphics systems to handle large numbers of moving models. For example, most contemporary flight simulators used Binary Space Partitioning, which is computationally effective for fixed environments since polygon display order (i.e., their depth coherence) can be pre-computed. While suitable for flight simulators, which largely have a point of view above the Earth's fixed surface, this technique is ineffective near the ground, where the order in which polygons overlay each other changes with the location of the point of view. It is also ineffective with a large number of moving models, since moving a model changes its depth coherence relative to the polygons representing the ground.

In contrast, Z-buffer techniques do not depend on pre-computed depth coherence. Therefore, it was a key enabling technology for SIMNET's on-ground point of view and its large numbers of moving vehicles. Z-buffering is memory intensive relative to Binary Space Partitioning but was made possible in part because the cost of RAM at the time had dropped significantly in price.

SIMNET used Z-buffer displays developed by Delta Graphics. Delta Graphics was founded by Drew Johnston (SW development), Mike Cyrus (President), both from the Boeing Aerospace Company/Graphics Lab, and Jay Beck (CTO and VP), a 3D graphics consultant of Softtool Consulting. The graphics processor, the GDP, custom developed for SIMNET by Gary Wilson (Sr. HW Engineer), won out over existing Silicon Graphics HW because of its low cost and because its architecture. It was the first simulator display processor to use a frame buffer and Z-buffer algorithms on a per display channel basis to show the simulated view.

Army Use of SIMNET for Training

The U.S. Army was actively used SIMNET for training, primarily at Fort Benning, Fort Rucker, and Fort Knox. Additional temporary and permanent locations were in Fort Leavenworth and Grafenwoehr, Germany.

SIMNET Follow-On Programs

The follow-on protocols to SIMNET were called Distributed Interactive Simulation; the primary U.S. Army follow-on program was the Close Combat Tactical Trainer (CCTT).

The SIMNET-D (Developmental) program used simulation systems developed in the SIMNET program to perform experiments in weapon systems, concepts, and tactics. It became the Advanced Simulation Technology Demonstration (ADST) program. It fostered the creation of the Battle Labs across the US Army, including the Mounted Warfare TestBed at

Ft Knox, Ky, the Soldier Battle Lab at Ft Benning, GA, the Air Maneuver Battle Lab at Ft Rucker, AL, the Fires Battle Lab at Ft Sill, OK.

Additional research programs after the end of SIMNET included work in weather and real-time terrain modifications.

The major software components of SIMNET have become almost standard fare among distributed simulators today (Strickland, Introduction, 2004).

- The Network Interface allows the software to interoperate with other simulators on a computer network.
- The Image Generator Software creates the beautiful full-color images that present the scenario to the training audience.
- A Controls & Displays Interface covert digital information from the computers into information that can be displayed on instruments in the vehicle. They also transform trainee input into digital signals that can be processed by the simulation computers.
- The Other-Vehicle State Table tracks the state data on the other vehicles in the scenario.
- Own Vehicle Dynamics software is used to model the vehicle in which the trainee is sitting. This software generates vehicle movement, firing, and communication.
- The Sound Generator recreates the deafening sounds of combat - the roar of tank engines, the clanking of tracks against the terrain, the firing of munitions, and the explosions of incoming rounds.

Modular Semi-Automated Forces

ModSAF is a set of software modules and applications that construct Distributed Interactive Simulation (DIS) and Computer Generated Forces (CGF) entities for realistic training, test, and evaluation on the virtual battlefield.

ModSAF was originally developed to provide a few dozen computer simulated 'opponents' for training exercises involving manned simulators. As simulation sizes increased, the nominal model of network message exchange fails. The HPC communications solution described in (Brunett & Gottschalk, 1997), with an explicit message router replacing network-based communications, enables ModSAF-style simulations with many tens of thousands of entities. Entity and environment models within ModSAF can be quite complex. The proceedings of the numerous DIS conferences, especially in the early 1990's, provide many demonstrations of the flexibility and utility of the ModSAF/DIS approach.

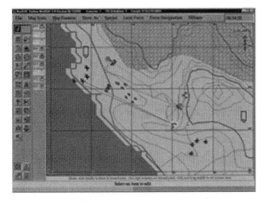

ModSAF:

- Was main DIS simulation library in use
- About 7000,000 lines of C
- Integer/logical intensive, relatively little floating point
- Irregular memory access patterns
- Broadcast communication between simulation nodes

Figure 2-2. Screenshot and characteristics of ModSAF

Description

ModSAF (Modular Semi-Automated Forces) is a set of software modules and applications used to construct Advanced Distributed Simulation (ADS) and Computer Generated Forces (CGF) applications. ModSAF modules and applications let a single operator create and control large numbers of entities that are used for realistic training, test, and evaluation on the virtual battlefield. ModSAF contains entities that are sufficiently realistic resulting in the user not being aware that computers, rather than human crews are maneuvering the displayed vehicles. These entities, which include ground and air vehicles, dismounted infantry (DI), missiles, and dynamic structures, can interact with each other and with manned individual entity simulators to support training, combat development experiments, and test of evaluation studies.

History

During the 1980's, the Defense Advanced Research Projects Agency (DARPA) developed virtual simulators for the M1 Abrams main battle tank and the M2 Bradley infantry-fighting vehicle to provide a sophisticated crew training capability. These simulators, each the replica of a single vehicle crew compartment, were then integrated on a local area network as a simulation network (SIMNET) to permit collective command and control (C2) and maneuver training of small units (armor platoon, etc.). SIMNET facilities were established in CONUS, USAREUR, and Korea.

As SIMNET users gained experience they needed a more robust and dynamic threat. DARPA therefore developed a constructive simulation to provide that opposing force (OPFOR) for users of SIMNET. This simulation used a semi-automated force (SAF) which could be task organized and given orders to execute basic military missions (attack, defend, etc.). This SIMNET SAF was an improvement but too inflexible. The SAF had been written in "hard code" and each time a change or improvement was made to the software, the SAF required 100 percent recompilation which required downtime and consumed man-hours and money.

DARPA therefore developed ModSAF. ModSAF meant, then as now, modular semi-automated forces. ModSAF version 1.0 was an improvement over SIMNET SAF because each major area of the software code (force characteristics, terrain, operating parameters, etc.) was now a separate module. Changes could be focused on the specific area requiring improvement and recompilation time was reduced.

Today, due to work by individual agencies on the different versions of ModSAF fielded for testing or use, there are many different variations available. Most of these versions have been developed by individual agencies to respond to agency-unique requirements. Users make changes to a version of ModSAF, as required, and the software modules in one user's version no longer resembles similar modules in another agency's version. The baseline versions (1.0, 2.0, and 3.0) of ModSAF exist under strict configuration control of STRICOM, the ModSAF material developer.

Capabilities for the planning process

See STOW ModSAF Site Operational Requirements (http://www-leav.army.mil/nsc/stow/saf/modsaf/oper.htm).

AIts capabilities with respect to planning in Verb/Noun Phrase format are:

ANALYZE:

- Outcome of simulation as a function of terrain, event or time
- Line-of-sight
- vehicle vulnerability
- probability of gunnery hits
- force strength
- intelligence
- logistics

- unit organization
- fire support missions
- After Action Review support
- Data reduction using add-on COTS/GOTS Tools

VISUALIZE:

- via a range of presentations tools and interfaces

ModSAF Use

ModSAF uses 'selective fidelity' to balance cost, desired performance, and realistic simulation. This concept emphasizes efficiency and avoids the simulation of behaviors and mechanisms that do not produce significant externally visible signatures. Therefore, many models include elements of human control that simplify the behavior of the entities. (ModSAF 5.0 Functional Description Document, ADST-II-CDRL-MODSAF5.0-9800327, ModSAF Web Site)

ModSAF is used within the SSEL by permission of the US DoD. Versions are hosted on both SGI IRIX systems and on PC platforms under Linux. It forms a fundamental pillar of the laboratory capability in teaching and demonstrating the principles of synthetic environments. It also provides a powerful enabler for the conduct of larger-scale academic SE exercises. As the development of Computer Generated Forces (CGF) and Semi Automated Forces (SAF) proceeds, increasingly it is also providing a base for active academic research in these areas. Recent examples of this have included MSc projects conducting validation exercises comparing CGF battles with live simulation exercises and research into Dismounted Infantry behaviors (an ongoing area of specialist research in the recently established Intelligent Systems and Virtual Environments Laboratory).

ONESAF

Description

One Semi-Automated Forces (OneSAF) is a next generation, entity-level simulation that supports both computer-generated forces and Semi-Automated Forces applications. This enables it to support a wide range of Army brigade-and-below constructive simulations and virtual simulators.

OneSAF is currently being integrated by the Synthetic Environment Core program as the replacement SAF for virtual trainers, such as Aviation Combined Arms Tactical Trainer and Close Combat Tactical Trainer (CCTT). OneSAF will serve as the basis for subsequent modernization activities for simulators across the Army. OneSAF was designed to represent the modular and Future Force and provides entities, units, and behaviors across the spectrum of military operations in the contemporary operating environment.

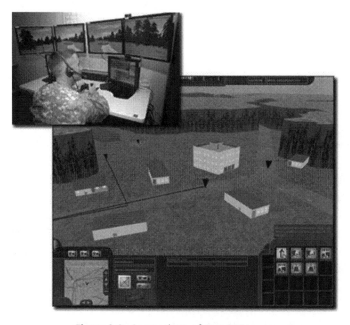

Figure 2-3. Screenshots of One SAF in use

OneSAF has been crafted to be uniquely capable of simulating aspects of the contemporary operating environment and its effects on simulated activities and behaviors. Special attention has been paid to detailed buildings for urban operations including interior rooms, furniture, tunnels and subterranean features, and associated automated behaviors to make use of these attributes. OneSAF is unique in its ability to model unit behaviors from fire team to company level for all units for both combat and non-combat operations. Intelligent, doctrinally correct behaviors and improved Graphical User Interfaces (GUIs) are provided to increase the span of control for workstation operators. The OneSAF Environmental Runtime Component (ERC) provides a range of terrain database services and capabilities already supporting live, virtual and constructive applications.

OneSAF represents a full range of operations, systems, and control processes in support of simulation applications applied to advanced concepts and requirements; research, development, and acquisition; and training, exercise, and military operations. OneSAF is designed to meet the constructive training challenges presented by transformation. With a full range of warfighter functional area representations, OneSAF displays a high fidelity environmental representation. OneSAF is a cross-domain simulation suitable for supporting training, analysis, research, experimentation, mission-planning, and rehearsal activities. It provides the latest physics-based modeling and data, enhanced data collection, and reporting capabilities.

In addition, interoperability support is present for industry standards such as Distributed Interactive Simulation (DIS), High Level Architecture (HLA), Military Scenario Development Language (MSDL), Joint Consultation Command and Control Information Exchange Data Model (JC3IEDM) and Army Battle Command System (ABCS) devices. OneSAF, as a cross-domain simulation suitable for supporting training, analysis, research, experimentation, mission-planning and rehearsal activities, provides the latest physics-based modeling and data, enhanced data collection and reporting capabilities.

Battlespace participants

The OneSAF agent architecture consists of actors, entities, and units. An actor is an entity or unit or a simulated thing that can be instantiated in a battlespace and has a location. An entity is the smallest discrete, stand-alone actor. It is implemented as a composition of physical components and behavioral components. A unit is an organized collection of actors and their capabilities or a collection of actors. In addition, it is simulation object representing the combined command and control of a collection of subordinate actors (entities and sub-units) or a collection of components (Logsdon, Nash, & Barnes, 2008).

The OneSAF Agent Architecture

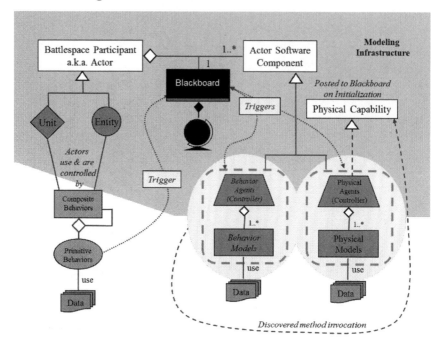

Figure 2-4. Depiction of the OneSAF Agent Architecture

Components

Components are agent/model sets. There are behavior and physical components for agents and models. Behavioral agents provide command and control capabilities, such as planning, plan execution, and situation assessment. Physical agents are the "middlemen" between behaviors, the physical world, and physical models.

Behavioral models answer behavior agents' questions and represent the reasoning of agents. Physical models provide physical capabilities, such as mobility, weapons, vulnerability, sensing, and communications. They represent the effectors and perceptors of simulated platforms and the physics of the simulated world.

A component has exactly one agent and one or more models. For example,

- The weapons controller uses one target selection model.
- The vulnerability agent uses multiple direct and indirect vulnerability models.

Table 2-1. Example Agents

Behavioral Agents:	Physical Agents:
<ul><li>Command</li><li>Scheduler</li><li>Intel</li><li>Message</li><li>Operations</li><li>Driver</li><li>Direct Fire Weapons Controller</li><li>Fire Direction Center</li><li>Caller For Fire</li><li>ADA Target Handoff</li></ul>	<ul><li>Weapon</li><li>Radio</li><li>Sensor</li><li>Mobility</li><li>Transport</li><li>Vulnerability</li></ul>

There are examples of components with only an agent, but these are poorly designed components. These components cannot be extended via composition (the OneSAF way!).

Behaviors

OneSAF behaviors include composite behaviors and primitive behaviors. Composite behaviors represent tasks and missions and are composed of primitive and other composite behaviors. They are created with the Behavior Composer.

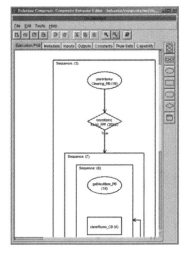

- The **Behavior Composer** uses meta-data to tie behaviors to the appropriate entities and units.
- The OneSAF Company and below C2 strategy Blocks A, B, and C uses Hierarchical Decomposition. Ordered behaviors are hierarchically decomposed at each echelon into trees with primitive behaviors at the leaf nodes and composite behaviors at the root and interior nodes.
- In general, to build behaviors:
  1. Write code for primitive behaviors.
  2. Use the Behavior Composer to build composite behaviors. The Behavior Composer allows placeholders for not yet written behaviors.

Figure 2-5. OneSAF behavior Composer

Primitive behaviors provide chunks of functionality from which more complex behavior models are built. They are parameterized with inputs, and may have outputs. They interact with behavioral agents.

State transitions

Within an actor, primitive behaviors and agents communicate via triggers posted on the blackboard with one exception. Using the blackboard decouples the behaviors and agents and fosters composability.

The exception is behavior agents directing physical agents. Instead of using triggers, each physical agent posts its capabilities on the blackboard at initialization. Each behavior agent retrieves the physical capabilities it needs from the blackboard at initialization. Then, during execution, the behavior agent directly uses the physical capabilities it has discovered during initialization.

Discovering and using physical capabilities still decouples the agents but is more efficient. This is the only allowed relatively direct connection between agents. Do NOT use the discovery of capabilities elsewhere (Logsdon, Nash, & Barnes, 2008).

In general, information outside of an actor reaches one or more (preferably one) components (usually an agent). The components process the information and generally posts information to the blackboard via a trigger. The blackboard treats triggers originating outside the actor just like triggers originating inside an actor.

For example, the radio component expressed interest in radio messages. When a radio message arrives, the appropriate message model is called and, if the message is received, the radio component creates a trigger for the message.

Another example is the vulnerability component expresses interest in detonation events. When a detonation arrives, the appropriate vulnerability models are called and, if there is damage, the vulnerability component creates a damage trigger.

Best practice example. Have only one component express interest in a simulation event and have that component post a trigger for other components. Occasionally, multiple components express interest in a simulation event. If that is the case, make sure the components post different triggers or check before creating duplicate triggers.

Content carriers

There are many simulation events, for example, fire and detonation events when shooting a weapon and collision events. Radio messages are a type of simulation event. Messages go through the communications architecture.

Directives are another type of simulation event. Directives are converted to triggers and posted to the blackboard. Agents are decoupled from directive issuers. Agents do not care who issued a directive. Be careful to not bind an agent to a directive sender. We want to be able to send directives from many places.

Facts are "information triggers." That is, facts like triggers are stored on the blackboard. Agents subscribe to facts like triggers. The difference is facts persist and when their values change, their subscribers are notified.

Triggers are intra-actor (that is, inter-agent) events. Agents do not address and send triggers to other agents. Instead, agents post and subscribe to triggers. If an agent posts a trigger that another agent subscribes to, the blackboard hands the trigger to the receiving agent. If multiple agents subscribe to a trigger, the blackboard gives the trigger to the agents in an arbitrary order. As a software developer, you will know that agent A is sending a trigger to agent B but do not encode that knowledge in the system. Also, do not assume an ordering of agent activation – OneSAF is composable, so you have no way of knowing what agents will be active in actors.

The Blackboard

The Blackboard is an agent control mechanism. Agents subscribe to triggers on the blackboard. When the blackboard receives a trigger, the it notifies that trigger's subscribers (and only those subscribers). The Blackboard interacts with "World Model" for an actor containing perceived truth. It sends and receives (internal) triggers, contains facts, and accesses physical capabilities. One instance per actor; cannot be shared with other actors. There is no worldModel object or instance. "World Model" is a term we use for discussing what an actor knows, believes, or perceives about itself and the simulated world (Logsdon, Nash, & Barnes, 2008). A blackboard instance exists in each actor.

Version 3.0

Table 2-2. Version 3.0 Enhancements

IED Capabilities	FCS
<ul><li>EFP (Explosively Formed Projectile)</li><li>Emplacement and Triggering</li><li>Detection</li><li>Disposal</li></ul>	<ul><li>Perception Forwarding</li><li>BLOS</li><li>NLOS</li></ul>
Aviation	**Miscellaneous**
	Data Collection
<ul><li>FARP</li><li>Fast rope, paratroop</li><li>MR Mobility Model</li></ul>	**MCT Updates** <ul><li>MSDL Import</li><li>Actor Attribute Manager</li></ul>
AMSAA Standard File Format	**Other**
Entities and Units	**User sponsored support**

Version 4.0

Table 2-3. Version 4.0 Enhancements

Physical Modeling	Combustible Fire
<ul><li>Visual Detection of IEDs (Subsurface, overpass, walls etc.)</li><li>Visual Detection data to reflect low probability of detection relative to other exposures of IEDs</li><li>IC Vulnerability to Fragmenting Munitions/IED</li></ul>	<ul><li>Ignition, Persistence, and Propagation of fire</li><li>Orderable behavior for Extinguishment of Fire</li><li>IC Vulnerability to fire</li></ul>
Behavioral Modeling	**Battle positions, Dig-in**
<ul><li>Detect and Report IED Construction materials</li><li>Detect, Report, and Secure area surrounding IEDs</li><li>Detect execution of IED emplacement, intervene to prevent</li><li>Analyze Explosion Site</li><li>Detect Covert Operations</li></ul>	<ul><li>Construct Earthen Fortifications</li><li>Occupy Fighting Position</li><li>Repair Damaged Terrain</li></ul>
	Contraband support added to
	<ul><li>Conduct Checkpoint</li><li>Conduct Prisoner Capture</li><li>Conduct RAID</li><li>Conduct Interview</li></ul>
	Data Loading
IED Visual representation based on enclosure and detonation status	<ul><li>Feedback mechanism to indicate all required data was loaded properly.</li><li>Feedback when data is not found or retrieved during runtime.</li></ul>

Chapter 3. Simulation Protocols/Architectures

The first widely applied technique for simulation interoperability was the SIMNET protocol that supported multiple, distributed tank simulators in a virtual environment. This evolved into the Distributed Interactive Simulation (DIS), which followed the same principles, but extended its application to serve virtual objects from all services and all service functions.

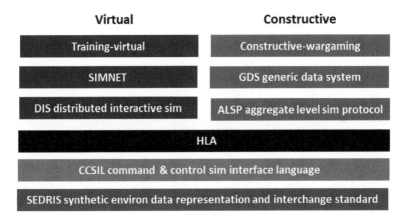

Figure 3-1. Different simulation protocol for virtual and constructive simulations, as well as federations that contain both types.

The Generic Data System (GDS) was developed at the Warrior Preparation Center in Germany. That center contained wargames from all of the services and was in desperate need of a method to make these work together. Around the same time MITRE began work on the Aggregate Level Simulation Protocol (ALSP) which became the global standard for tying constructive simulations together.

The High Level Architecture is a service-based technique that can be used for all types of simulations. The Command and Control Simulation Interface Language (CCSIL) is a standard way to encode command content for communication with C4I systems or between simulated commanders. The Synthetic Environment Data Interchange Standard (SEDRIS) is a method for achieving consistency between environmental databases in multiple system formats.

SIMNET

SIMNET is both a simulator (see Chapter 2) and a simulation protocol. The major software components of SIMNET have become almost standard fare among distributed simulators today. The Network Interface allows the software to interoperate with other simulators on a computer network. The Image Generator Software creates the beautiful full-color images that present the scenario to the training audience. A Controls & Displays Interface covert digital information from the computers into information that can be displayed on instruments in the vehicle. They also transform trainee input into digital signals that can be processed by the simulation computers. The Other-Vehicle State Table tracks the state data on the other vehicles in the scenario. Own Vehicle Dynamics software is used to model the vehicle in which the trainee is sitting. This software generates vehicle movement, firing, and communication. The Sound Generator recreates the deafening sounds of combat—the roar of tank engines, the clanking of tracks against the terrain, the firing of munitions, and the explosions of incoming rounds (Strickland, Introduction, 2004).

SIMNET

- Network Interface
- Image Generator Software
- Controls & Display Interface
- Other-Vehicle State Table
- Own-Vehicle Dynamics
- Sound Generation

Figure 3-2. Characteristics and screenshots of SIMNET

Distributed Interactive Simulation (DIS)

DIS is an extension of the concepts developed by the DARPA Simulation Networking (SIMNET) program. While SIMNET was a successful demonstration of homogenous simulation networking, it was recognized that a method for heterogeneous simulator networking was needed. A heterogeneous simulator network provides for the interaction of simulators developed by various contractors for disparate DoD organizations.

Distributed Interactive Simulation (DIS) is an open standard for conducting real-time platform-level wargaming across multiple host computers and is used worldwide, especially by military organizations but also by other agencies such as those involved in space exploration and medicine.

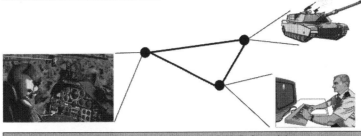

Distributed Interactive Simulation (DIS)

"The primary mission of DIS is to define an infrastructure for linking simulations of various types at multiple locations to create realistic, complex, virtual 'worlds' for the simulation of highly interactive activities" [DIS Vision, 1994].

- Developed in U.S. Department of Defense, initially for training
- DVEs widely used in DoD; growing use in other areas (entertainment, emergency planning, air traffic control)

Figure 3-3. Depiction of DIS with Live, Virtual, and Constructive simulation.

History

The standard was developed over a series of "DIS Workshops" at the Interactive Networked Simulation for Training symposium, held by the University of Central Florida's Institute for Simulation and Training (IST). The standard itself is very closely patterned after the original SIMNET distributed interactive simulation protocol, developed by Bolt, Beranek and Newman (BBN) for Defense Advanced Research Project Agency (DARPA) in the early through late 1980s. BBN introduced the concept of dead reckoning to efficiently transmit the state of battle field entities.

In the early 1990s, IST was contracted by the United States Defense Advanced Research Project Agency to undertake research in support of the US Army Simulator Network (SIMNET) program. Funding and research interest for DIS standards development decreased following the proposal and promulgation of its successor, the High Level Architecture (simulation) in 1996. HLA was produced by the merger of the DIS protocol with the Aggregate Level Simulation Protocol (ALSP) designed by MITRE.

There was a NATO standardization agreement (STANAG 4482, Standardized Information Technology Protocols for Distributed Interactive Simulation (DIS), adopted in 1995) on DIS for modeling and simulation interoperability. This was retired in favor of HLA in 1998 and officially cancelled in 2010 by the NATO Standardization Agency (NSA).

Protocol data units

[1278.1a-1998 IEEE Standard for Distributed Interactive Simulation - Application Protocols". IEEE. http://ieeexplore.ieee.org/servlet/opac?punumber =5896. Retrieved 10100517.]

The current version (DIS 6) defines 67 different PDU types, arranged into 12 families. Frequently used PDU types are listed below for each family. PDU and family names shown in italics are included in present draft DIS 7.

- Entity information/interaction family - Entity State, Collision, Collision-Elastic, Entity State Update, Attribute
- Warfare family - Fire, Detonation, Directed Energy Fire, Entity Damage Status
- Logistics family - Service Request, Resupply Offer, Resupply Received, Resupply Cancel, Repair Complete, Repair Response
- Simulation management family - Start/Resume, Stop/Freeze, Acknowledge
- Distributed emission regeneration family - Designator, Electromagnetic Emission, IFF/ATC/NAVAIDS, Underwater Acoustic, Supplemental Emission/Entity State (SEES)
- Radio communications family - Transmitter, Signal, Receiver, Intercom Signal, Intercom Control
- Entity management family
- Minefield family
- Synthetic environment family
- Simulation management with reliability family

- Live entity family
- Non-real time family
- Information Operations family - Information Operations Action, Information Operations Report

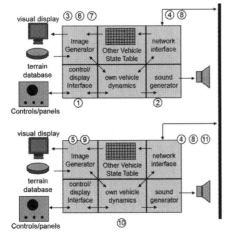

Typical Sequence

1. Detect trigger press
2. Audio "fire" sound
3. Display muzzle flash
4. Send fire PDU
5. Display muzzle flash
6. Compute trajectory, display tracer
7. Display shell impact
8. Send detonation PDU
9. Display shell impact
10. Compute damage
11. Send Entity state PDU indicating damage

Figure 3-4. Example of how DIS works

Distributed Interactive Simulation (DIS) provides an environment where military manned simulators, semi-autonomous forces, and live, instrumented equipment can communicate over the Distributed Simulation Internet (DSI) and create a virtual battlefield. DIS works by communicating information via Protocol Data Units (PDUs). PDUs are broadcast across the DSI and interpreted by the simulators or equipment on the other end. Based on the messages in the PDUs, each simulator's Computer Image Generator (CIG) renders an image of the other vehicle if it is in its own field of view. In addition to the image of the other simulators, PDUs register simulation events, such as firing munitions. Based on internal models, if the munitions hit a vehicle, the vehicle will process the information and determine the effect of the damage.

Table 3-1. Basic DIS Concepts

• Autonomous simulation nodes responsible for maintaining the state of simulation entities.
• No central computer for event scheduling or conflict resolution.
• Use a standard protocol for communicating "ground truth" data
• Receiving nodes are responsible for determining what is perceived
• Simulation nodes communicate primarily changes of the state of their entities.
    o Dead reckoning to reduce communications.
    o Typical message rate is 5 times a second per entity

The prime design concept of DIS is that each simulator node is autonomous and simulates a single battlespace entity, or group of entities in the case of semi-automated forces (SAF) systems. DIS supports the distributed concept of simulation, in which each simulator node in the configuration supplies all the resources necessary for its own processing.

Another design principle of DIS is that, within a particular configuration, there is no central computer with the responsibility for interaction detection and resolution. This prevents single-point failures from disrupting the overall exercise or mission being supported by a DIS configuration. Each node is responsible for determining its own interaction with the rest of the electronic battlespace. A node communicates any changes in state caused by interactions and resolutions to the rest of the configuration.

To communicate its current status (location, velocity, orientation, etc.) to all other entities within a particular exercise, each entity's host computer broadcasts updates to its status. The receiving entity's host computer takes this status information from all the other entities in the exercise and calculates ground truth. The computer can then determine what can be seen by the entity it is simulating and updates displays as necessary.

DIS reduces communication requirements by requiring that each node maintain a simple model for all of the entities on the battlespace that it is not simulating. These entities are called remote entities. Between the receipt of updates from the remote nodes actually responsible for simulating those entities, the local node executes these simple dead reckoning models for those entities to estimate their current state.

Distributed Wargaming System

The Distributed Wargaming System concept was conceived and pioneered at the Center for use in the first ever NATO-wide simulation exercise Allied Command Europe (ACE 89). This effort grew until Global Distributed Wargaming became a reality. In 1992, the WPC began to assist the Korea Battle Simulation Center with the conduct of the annual exercise Ulchi Focus Lens. More recently, the Warrior Preparation Center (WPC) has set up and hosted numerous exercises with worldwide focus. In May 1999, Operation Shining Hope kicked off with the WPC providing administrative and logistical support. President Clinton visited the WPC to observe the exercise in progress. In January 2000, the WPC hosted EC00, a USCINCEUR-directed, USAREUR-led Joint Task Force training exercise consisting of over 600 participants with training goals of rehearsing and refining operations, as well as activating and deploying a Joint Task Force. In November 2000, Constant Harmony 00, a distributed NATO Joint Warfighter exercise was conducted at the WPC for the Regional Headquarters Allied Forces North Europe and its staff. The USAFE sponsored exercise Union Flash 01 was conducted at the WPC and Ramstein Air Base in April 2001, training the Joint Force Air Component Commander, The Joint Air Operations Center and the USAREUR Battlefield Coordination Element.

A common distributed wargaming system is the Joint Training Confederation (JTC). The Department of Defense has a long history of using simulations to train soldiers to perform missions and to learn to work together in teams and across command structures. In the area of command and staff training these simulations have typically been interactive, constructive, time-stepped models. Several of these have emerged as the most prominent tools for service specific and Joint training. In order to bring all of the services into a single synthetic environment for Joint operations similar to those experienced during Desert Storm, these models have been joined into a confederation.

The JTC was created from existing models that were not intended to operate together. Each was built to serve a specific customer, focusing the fidelity on the functions that were most needed for the intended customer. The confederation of these models allow all of the services to train together using the model that best serves their needs, but to do so in an environment that is synchronized with the other service models. The creation of this confederation has been very effective, but it has illustrated the difficulties of joining simulations that contain fundamentally different assumptions about the way combat should be modeled. As a result, compromises were arrived at to make the JTC operate.

The models in the JTC have been built, enhanced, and integrated over the past 15 years and are reaching the end of their service lives. As the simulation community considers the creation of the next generation of models to replace these, it is clear that interoperability among them is a primary requirement. In constructing the JTC, two characteristics came to light. The first was that each service had built a large body of software that were duplicates of each other. Each contained functions to manage time, calculate location, manage lists of entities, provide user interfaces, and a host of other services. This duplication presented an opportunity for the application of software reuse techniques across the models. Second, the shadows of a fundamental architecture could be seen running through each of the systems. Each contained tools for scenario generation, after action analysis, data storage, trainee interfaces, controller interfaces, and an engine for the execution of simulation events. The Joint Training Confederation is composed of nine simulations or actors (Griffin, Page, Furness, & Fischer, 1997).

- Air Warfare Simulation (AWSIM)
- Corps Battle Simulation (CBS)
- Joint Command and Control Warfare Simulation System (JCCWSS) JQUAD, which is an ALSP actor with four simulations: Joint Electronic Combat-Electronic Warfare Simulation (JECEWSI), Joint Command and Control Attack Simulation (JCAS), Joint Operations Information Simulation (JOISIM), and Joint Networks Simulation (JNETS).
- MAGTF Tactical Warfare Simulation (MTWS)
- Research, Evaluation and Systems Analysis (RESA) Simulation
- Tactical Simulation Model (TACSIM)
- Combat Service Support Training Simulation System (CSSTSS)
- Missile Defense Space Tool (MDST)

Joint Training Confederation

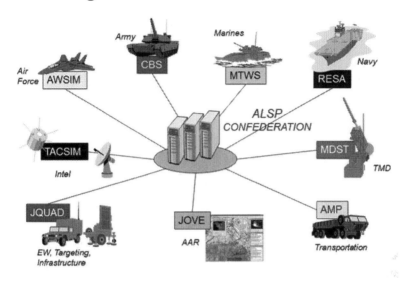

Figure 3-5. Components of the Joint Training Confederation (JTC)

JTC actors are interconnected with the AIS. Possible configurations of the JTC actors and the infrastructure software are shown here. This figure highlights:

- Each JTC actor has a translator that converts ALSP formatted messages into a form compatible with that specific simulation.
- The ALSP software consists of four components: the ALSP Common Module (ACM), the ALSP Broadcast Emulator (ABE) and the ALSP Controller Terminal (ACT) and the ALSP Confederation Management Tool (CMT).
- The AIS components communicate with each other through a network of local area networks (LANs) and/or wide area networks (WANs).

Aggregate Level Simulation Protocol (ASLP)

Aggregate Level Simulation Protocol (ALSP), both software and a protocol, is used to enable disparate simulations to communicate with one another. It is used extensively by the United States military to link analytic and training simulations to support training requirements for Corps and above. It has been for wargames, what DIS has been for virtual simulators.

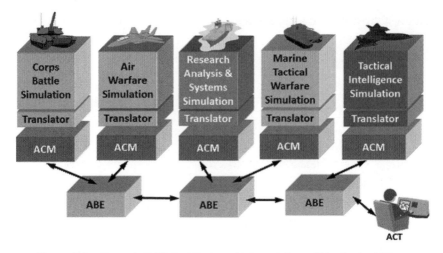

Figure 3-6. Current ALPS Architecture (Adapted from (Weatherly, Wilson, Canova, Page, Zabek, & Fischer, 1995)

ALSP consists of three components: (1) participating simulations or actors adapted for use with ALSP; (2) the ALSP Infrastructure Software (AIS) providing distributed runtime simulation support and management; (3) a reusable ALSP Interface consisting of a set of generic data exchange message protocols (i.e., formal rules for information exchange) to enable interaction among objects represented in different simulations.

History

In 1990, the Defense Advanced Research Projects Agency (DARPA) employed The MITRE Corporation to study the application of distributed interactive simulation principles employed in SIMNET to aggregate-level constructive training simulations. Based on prototype efforts, a community-based experiment was conducted in 1991 to extend SIMNET to link the US Army's Corps Battle Simulation (CBS) and the US Air Force's Air Warfare Simulation (AWSIM). The success of the prototype and users' recognition of the value of this technology to the training community led to development of production software. The first ALSP confederation, providing air-ground interactions between CBS and AWSIM, supported three major exercises in 1992.

By 1995, ALSP had transitioned to a multi-Service program with simulations representing the US Army (CBS), the US Air Force (AWSIM), the US Navy (RESA), the US Marine Corps (MTWS), electronic warfare (JECEWSI), logistics (CSSTSS), and intelligence (TACSIM). The program had also transitioned from DARPA's research and development emphasis to mainstream management by the US Army's Program Executive Office for Simulation, Training, and Instrumentation (PEO STRI).

Contributions

ALSP developed and demonstrated key aspects of distributed simulation, many of which were applied in the development of HLA.

- No central node so that simulations can join and depart from the confederation at will
- Geographic distribution where simulators can be distributed to different geographic locations yet exercise in the same simulated environment
- Object ownership so each simulation controls its own resources, fires its own weapons and determines appropriate damage to its systems when fired upon
- A message-based protocol for distributing information from one simulation to all other simulations.
- Time management so that the times for all simulations appear the same to users and so that event causality is maintained—events should occur in the same sequence in all simulations.
- Data management permits all simulations to share information in a commonly understood manner even though each had its own representation of data. This includes multiple simulations controlling attributes of the same object.
- An architecture that permits simulations to continue to use their existing architectures while participating in an ALSP confederation.

Motivation

In 1989, the Warrior Preparation Center (WPC) in Einsiedlerhof, Germany hosted the computerized military exercise ACE-89. The Defense Advanced Research Projects Agency (DARPA) used ACE-89 as a technology insertion opportunity by funding deployment of the Defense Simulation Internet (DSI). Its packetized video teleconferencing brought general officers of NATO nations face-to-face during a military exercise for the first time; this was well-received. But the software application of DSI, distribution of Ground Warfare Simulation (GRWSIM), was less successful. The GRWSIM simulation was unreliable and its distributed database was inconsistent, degrading the effectiveness of the exercise (Weatherly, Wilson, Canova, Page, Zabek, & Fischer, 1995).

DARPA was funding development of a distributed tank trainer system called SIMNET where individual, computerized, tank-crew trainers were connected over local area networks and the DSI to cooperate in a single, virtual battlefield. The success of SIMNET, the disappointment of ACE-89, and the desire to combine existing combat simulations prompted DARPA to initiate research that lead to ALSP.

Basic Tenets

DARPA sponsored the design of a general interface between large, existing, aggregate-level combat simulations. Aggregate-level combat simulations use Lanchestrian models of combat rather than individual physical weapon models and are typically used for high-level training. Despite representational differences, several principles of SIMNET applied to aggregate-level simulations (Strickland, Introduction, 2004):

- **Dynamic configurability.** Simulations may join and depart an exercise without restriction.
- **Geographic distribution.** Simulations can reside in different geographic locations yet exercise over the same logical terrain.
- **Autonomous entities.** Each simulation controls its own resources, fires its own weapons and, when one of its objects is hit, conducts damage assessment locally.
- **Communication by message passing.** A simulation uses a message-passing protocol distribute information to all other simulations.

The ALSP challenge had requirements beyond those of SIMNET (Weatherly, Wilson, Canova, Page, Zabek, & Fischer, 1995):

- **Simulation time management.** Typically, simulation time is independent of wall-clock time. For the results of a distributed simulation to be "correct," time must be consistent across all simulations (Lamport, 1978).
- **Data management.** The schemes for internal state representation differ among existing simulations, necessitating a common representational system and concomitant mapping and control mechanisms.

- **Architecture independence.** Architectural characteristics (implementation language, user interface, and time flow mechanism) of existing simulations differed. The architecture implied by ALSP must be unobtrusive to existing architectures.

Conceptual Framework

A conceptual framework is an organizing structure of concepts that facilitates simulation model development (Balci, Nance, Derrick, Page, & Bishop, 1990). Common conceptual frameworks include: event scheduling, activity scanning and process interaction.

The ALSP conceptual framework is object-based where a model is composed of objects that are characterized by attributes to which values are assigned. Object classes are organized hierarchically in much the same manner as with object-oriented programming languages. ALSP supports a confederation of simulations that coordinate using a common model.

To design a mechanism that permits existing simulations to interact, two strategies are possible: (1) define an infrastructure that translates between the representations in each simulation, or (2) define a common representational scheme and require all simulations to map to that scheme.

The first strategy requires few perturbations to existing simulations; interaction is facilitated entirely through the interconnection infrastructure. However, this solution does not scale well. Because of an underlying requirement for scalability, the ALSP design adopted the second strategy. ALSP prescribes that each simulation maps between the representational scheme of the confederation and its own representational scheme. This mapping represents one of the three ways in which a simulation must be altered to participate in an ALSP confederation. The remaining modifications are (Weatherly, Wilson, Canova, Page, Zabek, & Fischer, 1995):

- Recognizing that the simulation doesn't own all of the objects that it perceives.
- Modifying the simulation's internal time advance mechanism so that it works cooperatively with the other simulations within the confederation.

In stand-alone simulations, objects come into (and go out of) existence with the passage of simulation time and the disposition of these objects is solely the purview of the simulation. When acting within a confederation, the simulation-object relationship is more complicated.

The simulation-object ownership property is dynamic, i.e. during its lifetime an object may be owned by more than one simulation. In fact, for any value of simulation time, several simulations may own different attributes of a given object. By convention, a simulation owns an object if it owns the "identifying" attribute of the object. Owning an object's attribute means that a simulation is responsible for calculating and reporting changes to the value of the attribute. Objects not owned by a particular simulation but within the area of perception for the simulation are known as ghosts. Ghosts are local copies of objects owned by other simulations.

When a simulation creates an object, it reports this fact to the confederation to let other simulations create ghosts. Likewise, when a simulation deletes an object, it reports this fact to enable ghost deletion. Whenever a simulation takes an action between one of its objects and a ghost, the simulation must report this to the confederation. In the parlance of ALSP, this is an interaction. These fundamental concepts provide the basis for the remainder of the presentation. The term confederation model describes the object hierarchy, attributes and interactions supported by a confederation.

ALSP Infrastructure Software (AIS)

The object-based conceptual framework adopted by ALSP defines classes of information that must be distributed. The ALSP Infrastructure Software (AIS) provides data distribution and process coordination. Principal components of AIS are the ALSP Common Module (ACM) and the ALSP Broadcast Emulator (ABE) (Weatherly, Wilson, Canova, Page, Zabek, & Fischer, 1995).

ALSP Common Module (ACM)

The ALSP Common Module (ACM) provides a common interface for all simulations and contains the essential functionality for ALSP. One ACM instance exists for each simulation in a confederation. ACM services require time management and object management; they include (Weatherly, Wilson, Canova, Page, Zabek, & Fischer, 1995):

- Coordinate simulations joining and departing from a confederation..
- Coordinate simulation local time with confederation time.
- Filter incoming messages, so that simulations receive only messages of interest.
- Coordinate ownership of object attributes, and permit ownership migration.
- Enforce attribute ownership so that simulations report values only for attributes they own.

Time management

Joining and departing a confederation is an integral part of time management process. When a simulation joins a confederation, all other ACMs in the confederation create input message queues for the new simulation. Conversely, when a simulation departs a confederation the other ACMs delete input message queues for that simulation (Weatherly, Wilson, Canova, Page, Zabek, & Fischer, 1995).

ALSP time management facilities support discrete event simulation using either asynchronous (next-event) or synchronous (time-stepped) time advance mechanisms (Nance, 1971). The mechanism to support next-event simulations is

1. A simulation sends an event-request message to its ACM with a time parameter corresponding to simulation time T, (the time of its next local event).
2. If the ACM has messages for its simulation with timestamps older than or the same as T, the ACM sends the oldest one to the simulation. If all messages have timestamps newer than T, the ACM send a grant-advance to the simulation, giving it permission to process its local event at time T.
3. The simulation sends any messages resulting from the event to its ACM.
4. The simulation repeats from step (1).

The mechanism to support time-stepped simulation is:

1. The simulation processes all events for some time interval $(T, T + \Delta T]$.
2. The simulation sends an advance request to its ACM for time $T + \Delta T$.

3. The ACM sends all messages with time stamps on the interval $(T, T + \Delta T]$ to the simulation, followed by a grant-advance to $T + \Delta T$.
4. The simulation sends any messages for the interval $(T, T + \Delta T]$ to the ACM.
5. The simulation repeats from step (1).

AIS includes a deadlock avoidance mechanism using null messages. The mechanism requires that the processes have exploitable look-ahead characteristics.

Object management

The ACM administers attribute database and filter information. The attribute database maintains objects known to the simulation, either owned or ghosted, and attributes of those objects that the simulation currently owns. For any object class, attributes may be members of (Weatherly, Wilson, Canova, Page, Zabek, & Fischer, 1995)

- **Create set.** Attributes minimally required to represent an object
- **Interest set.** Useful, but not mandatory, information
- **Update set.** Object attribute values reported by a simulation to the confederation

Information flow across the network can be further restricted through filters. Filtering provides discrimination by (1) object class, (2) attribute value or range, and (3) geographic location. Filters also define the interactions relevant to a simulation.

```
If (an update passes all filter criteria)
|  If (the object is known to the simulation)
|  |  Send new attribute values to simulation
|  Else (object is unknown)
|  |  If (enough information is present to create a ghost)
|  |  |  Send a create message to the simulation
|  |  Else (not enough information is known)
|  |  |  Store information provided
|  |  |  Send a request to the confederation for missing data
Else (the update fails filter criteria)
|  If (the object is known to the simulation)
|  |  Send a delete message to the simulation
|  Else
|  |  Discard the update data
```

The ownership and filtering information maintained by the ACM provide the information necessary to coordinate the transfer of attribute ownership between simulations.

ALSP Broadcast Emulator (ABE)

An ALSP Broadcast Emulator (ABE) facilitates the distribution of ALSP information. It receives a message on one of its communications paths and retransmits the message on all of its remaining communications paths. This permits configurations where all ALSP components are local to one another (on the same computer or on a local area network). It also permits configurations where sets of ACMs communicate with their own local ABE with inter-ABE communication over wide area networks (Weatherly, Wilson, Canova, Page, Zabek, & Fischer, 1995).

HLA Overview

The **High Level Architecture (HLA)** is a general purpose architecture for distributed computer simulation systems. Using HLA, computer simulations can interact (that is, to communicate data, and to synchronize actions) to other computer simulations regardless of the computing platforms. A Run-Time Infrastructure (RTI) manages the interaction between simulations (Kuhl, Weatherly, & Dahmann, 1999).

The **HLA** was designed to support all training simulations, virtual or constructive. It replaces SIMNET, DIS, ALSP, and other custom protocols.
 HLA Compliance
 • HLA Rules (10 rules)
   o Defines the interoperation of federates (simulations) in a federation (family of simulations).
 • Interface Specification (RTI software)
   o Exchange, send, receive, data from remote simulators in a standardized way.
 • Object Model Template
   o Describe capabilities of a simulator

Figure 3-7. Three main parts i=of the High-Level Architecture (HLA)

Technical overview

The High Level Architecture (HLA) is an architecture for reuse and interoperation of simulations. The HLA is based on the premise that no single simulation can satisfy the requirements of all uses and users. An individual simulation or set of simulations developed for one purpose can be applied to another application under the HLA concept of the federation: a composable set of interacting simulations. The intent of the HLA is to provide a structure that will support reuse of capabilities available in different simulations, ultimately reducing the cost and time required to create a synthetic environment for a new purpose and providing developers the option of distributed collaborative development of complex simulation applications.

The HLA was developed under the leadership of the Defense Modeling and Simulation Office (DMSO) to support reuse and interoperability across the large numbers of different types of simulations developed and maintained by the DoD. The HLA was approved as an open standard through the Institute of Electrical and Electronic Engineers (IEEE) - IEEE Standard 1516 - in September 2000. In November 2000, the Services and Joint Staff signed the HLA Memorandum of Agreement identifying the HLA as the preferred architecture for simulation interoperability within the DoD.

A High Level Architecture consists of the following components:

- Interface Specification, that defines how HLA compliant simulators interact with the Run-Time Infrastructure (RTI). The RTI provides a programming library and an application-programming interface (API) compliant to the interface specification.
- Object Model Template (OMT) that specifies what information is communicated between simulations, and how it is documented.
- Rules, that simulation must obey in order to be compliant to the standard.

Common HLA terminology

- Federate: an HLA compliant simulation entity.
- Federation: multiple simulation entities connected via the RTI using a common OMT.
- Object: a collection of related data sent between simulations.
- Attribute: data field of an object.

- Interaction: event sent between simulation entities.
- Parameter: data field of an interaction.

Interface specification

The interface specification is object oriented. Many RTIs provide APIs in C++ and the Java programming languages. While the HLA is an architecture, not software, use of runtime infrastructure (RTI) software is required to support operations of a federation execution. The RTI software provides a set of services used by federates to coordinate their operations and data exchange during a runtime execution. Access to these services is defined by the HLA Interface Specification.

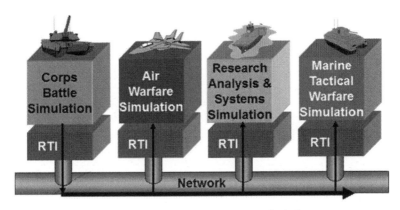

Figure 3-8. Runtime Infrastructure (RTI) of a federation

The interface specification is divided into service groups:

- Federation Management
- Declaration Management
- Object Management
- Ownership Management
- Time Management
- Data Distribution Management
- Support Services

Object model template

The object model template (OMT) provides a common framework for the communication between HLA simulations. OMT consists of the following documents:

- **Federation Object Model** (FOM). The FOM describes the shared object, attributes and interactions for the whole federation.
- **Simulation Object Model** (SOM). A SOM describes the shared object, attributes and interactions used for a single federate.

Reusability and interoperability require that all objects and interactions managed by a federate, and visible outside the federate, should be specified in detail and with a common format. The Object Model Template (OMT) provides a standard for documenting HLA Object Model information. The OMT defines the Federation Object Model (FOM), the Simulation (Federate) Object Model (SOM), and the Management Object Model (MOM).

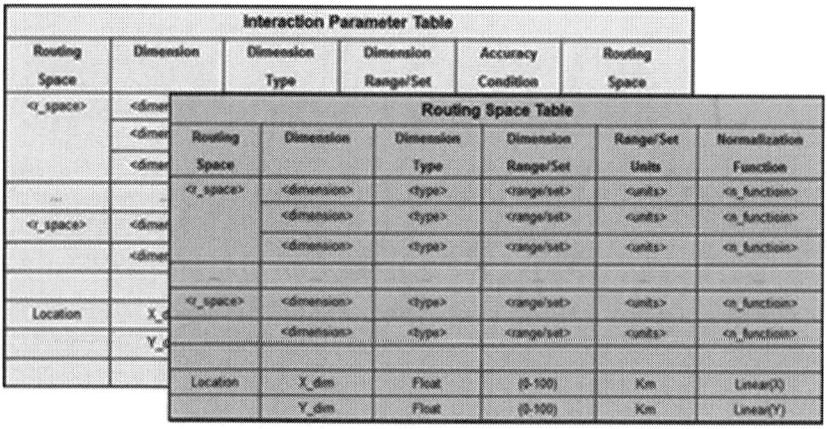

Figure 3-9. Object model template examples

HLA rules

The HLA rules describe the responsibilities of federations and the federates that join (DMSO, 2001).

- Federations shall have an HLA Federation Object Model (FOM), documented in accordance with the HLA Object Model Template (OMT).
- In a federation, all representation of objects in the FOM shall be in the federates, not in the run-time infrastructure (RTI).
- During a federation execution, all exchange of FOM data among federates shall occur via the RTI.

- During a federation execution, federates shall interact with the run-time infrastructure (RTI) in accordance with the HLA interface specification.
- During a federation execution, an attribute of an instance of an object shall be owned by only one federate at any given time.
- Federates shall have an HLA Simulation Object Model (SOM), documented in accordance with the HLA Object Model Template (OMT).
- Federates shall be able to update and/or reflect any attributes of objects in their SOM and send and/or receive SOM object interactions externally, as specified in their SOM.
- Federates shall be able to transfer and/or accept ownership of an attribute dynamically during a federation execution, as specified in their SOM.
- Federates shall be able to vary the conditions under which they provide updates of attributes of objects, as specified in their SOM.
- Federates shall be able to manage local time in a way that will allow them to coordinate data exchange with other members of a federation.

Base Object Model

The Base Object Model (BOM) is a new concept created by SISO to provide better reuse and composability for HLA simulations, and is highly relevant for HLA developers. More information can be found at Boms.info.

Federation Development and Execution Process (FEDEP)

FEDEP, IEEE 1516.3-2003, is a standardized and recommended process for developing interoperable HLA based federations. FEDEP is an overall framework overlay that can be used together with many other, commonly used development methodologies.

Distributed Simulation Engineering and Execution Process (DSEEP)

In spring 2007 SISO started revising the FEDEP. It has been renamed to Distributed Simulation Engineering and Execution Process (DSEEP) and is now an active standard IEEE 1730-2010 (instead of IEEE 1516.3).

Standards

HLA is defined under IEEE Standard 1516:

- IEEE 1516-2010 - Standard for Modeling and Simulation High Level Architecture - Framework and Rules
- IEEE 1516.1-2010 - Standard for Modeling and Simulation High Level Architecture - Federate Interface Specification
- IEEE 1516.2-2010 - Standard for Modeling and Simulation High Level Architecture - Object Model Template (OMT) Specification
- IEEE 1516.3-2003 - Recommended Practice for High Level Architecture Federation Development and Execution Process (FEDEP)
- IEEE 1516.4-2007 - Recommended Practice for Verification, Validation, and Accreditation of a Federation an Overlay to the High Level Architecture Federation Development and Execution Process

Machine-readable parts of the standard, such as XML Schemas, C++, Java and WSDL APIs as well as FOM/SOM samples can be downloaded from the IEEE 1516 download area of the IEEE web site. The full standards texts are available at no extra cost to SISO members or can be purchased from the IEEE shop.

Prior to publication of IEEE 1516, the US Defense Modeling and Simulation Office sponsored the HLA standards development. The first complete version of the standard, published 1998, was known as HLA 1.3.

STANAG 4603

HLA (in both the current IEEE 1516 version and its ancestor "1.3" version) is the subject of the NATO standardization agreement (STANAG 4603) for modeling and simulation: Modeling And Simulation Architecture Standards For Technical Interoperability: High Level Architecture (HLA).

DLC API

SISO has developed a complementary HLA API specification known as the Dynamic Link Compatible (DLC) API. The DLC API addresses a limitation of the IEEE 1516 and 1.3 API specification, whereby federate recompilation was necessary for each different RTI implementation. Note that this API has since been superseded by the HLA Evolved APIs, informally known as Evolved DLC APIs (EDLC).

- SISO-STD-004-2004 - Dynamic Link Compatible HLA API Standard for the HLA Interface Specification Version 1.3

- SISO-STD-004.1-2004 - Dynamic Link Compatible HLA API Standard for the HLA Interface Specification (IEEE 1516.1 Version)

Computer Wargame Evolution

Modern wargaming originated with the military need to study warfare and to 'reenact' old battles for instructional purposes. The stunning Prussian victory over the Second French Empire in the Franco-Prussian War of 1870-1871 is sometimes partly credited to the training of Prussian officers with the game *Kriegspiel*, which was invented around 1811 and gained popularity with many officers in the Prussian army. These first wargames were played with dice which represented "friction", or the intrusion of less than ideal circumstances during a real war (including morale, weather, the fog of war, etc.), though this was usually replaced by an umpire who used his own combat experience to determine the results (Leeson, 2011).

The Cold War provided fuel for many computer wargames that attempted to show what a non-nuclear (or, in a very few cases, nuclear) World War III would be like, moving from a re-creation to a predictive model in the process. The figure below shows the model genealogy beginning in the early 1970's and leading up to the abandoned development of the members of the Joint Simulation System (JSIMS). The Joint Training Confederation (JTC) was an interoperability program that was intended to join together several models that had originally been designed to operate independently. JSIMS was attempting to design the entire family to operate together from the beginning.

Computer Wargame Evolution

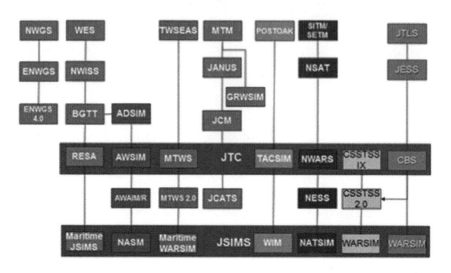

Figure 3-10. Depiction of the evolution of computer wargames.

Joining models after they are created has proven to provide only a very limited degree of interoperability. Each model has a specific representation of the world that allows it to share/export information in very limited ways. However, the JTC program proved that interoperability at this level was not feasible.

Table 3-2. Some "Well-Known" Simulations

	Primary Domain	Status	Primary Role	Scope	Aggregation	Stochastic?	Time Cont
TACWAR	Joint	In use	Opns	Theater	Bn/Bde	Determ.	T-Step
JWARS[1]	Joint	In Devel	Analy/ Acq	Campaign	Company	Stoch.	Event
CEM	Ground	In use	Analy	Theater	Bde/Div	Determ.	T-Step
JSIMS	Joint	In Devel	Tng	Theater	Variable	Stoch.	Event
VIC	Ground	In Use	Analy	Corps/Div	Battalion	Determ.	T-Step
Eagle	Ground	In Use	An/ Tng	Corps/Div	Company	Determ.	T-Step
CASTFOREM	Ground	In Use	Analy	Bn/Bde	Entity	Stoch.	Event
COMBAT XXI[2]	Ground	In Devel	Analy	Bn/Bde	Entity	Stoch.	Event/ Agent
JANUS	Ground	In Use	An/ Tng	Bn/Bde	Entity	Stoch.	Event
ModSAF	Ground	In Use	Tng	Bn/Bde	Entity	Stoch.	Event
One SAF[3]	Ground	In Use	Analy/ Tng/ Exp	Crew/Bde	Entity	Stoch.	Event
CBS	Ground	In Use	Tng	Corps	Battalion	Stoch.	T-Step
WARSIM[4]	Ground	In Use	Analy/ Tng	Corps	Company	Stoch.	Event
NSS	Naval/Air	In Use	Analy	Daily Ops	Entity	Stoch.	Event
EADSIM	Air/AD	In Use	Analy	Daily Ops	Entity	Stoch.	T-Step
THUNDER	Air	In Use	Analy	Campaign	Entity	Stoch.	T-Step
JICM	Global	In Use	Analy	Global	Bde/Div	Determ.	T-Step
EPiCS	Ground	In Use	Analy	Emerg	Bde/Bn	Stoch.	

Tng = training; Analy = analysis; Acq = acquisition; Stoch = stochastic; Determ = deterministic; Event = discrete event; T-Step = time step

---

[1] Replacement for TACWAR
[2] Replacement for CASTFOREM
[3] Replacement for ModSAF
[4] Replacement for CBS

Verification, Validation and Accreditation

The users of a simulation must have confidence that the simulation results are meaningful and that the simulation output is representative of actual missile performance. It is essential that the models of the missile system, subsystems, and physical environment have a demonstrable correspondence with the system, subsystem, or environment being modeled. This confidence is gained through the processes of verification and validation. Verification ensures that the computer program operates correctly according to the conceptual model of the missile system. Validation determines the extent to which the simulation is an accurate representation of the real world (Strickland, Missile Flight Simulation: Surface-to-Air Missiles, 2010).

In a simulation, the real-world system is abstracted by a conceptual model—a series of mathematical and logical relationships concerning the components and structure of the system. The conceptual model is then coded into a computer recognizable form (i.e., an operational model), which we hope is an accurate imitation of the real-world system. The accuracy of the simulation must be checked before we can make valid conclusions based on the results from a number of runs. This checking process consists of three main components:

- **Verification**: The process of determining that a model implementation and its associated data accurately represent the developer's conceptual description and specifications.
- **Validation**: The process of determining the degree to which a model and its associated data provide an accurate representation of the real world from the perspective of the intended uses of the model.
- **Accreditation**: The official certification that a model, simulation, or federation of models and simulations and its associated data is acceptable for use for a specific purpose. [DoDI 5000.61]

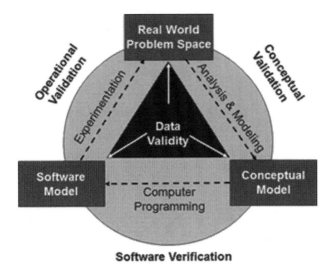

Figure 3-11. A depiction of the V&A process

It can also be helpful to remember each one in terms of simple question that (informally) captures the essential idea:

- **Verification** – Did I build the thing right?
- **Validation** – Did I build the right thing?
- **Accreditation** – Should it be used?

Also, there is an underlying implicit principle, and its key question:

- **Credibility** – Should it be trusted?

An accreditation decision reflects a determination that the evidence supporting a decision on whether and "how" to employ a simulation is strong enough to warrant putting that conclusion in writing and creating an official record of the decision – something not to be taken lightly.

Netcentric Warfare

"We must build forces that draw upon the revolutionary advances in the technology of war…one that relies more heavily on stealth, precision weaponry, and information technologies."

George W. Bush,
Commander in Chief

Network Centric Warfare (NCW) or Netcentric Warfare is no less than the embodiment of an Information Age transformation of the DoD. It involves a new way of thinking about how we accomplish our missions, how we organize and interrelate, and how we acquire and field the systems that support us. NCW moves the Department to the next level of Jointness as envisioned in Joint Vision 2020. This monumental task will span a quarter century or more. It will involve ways of operating that have yet to be conceived, and will employ technologies yet to be invented. NCW has the potential to increase warfighting capabilities by orders of magnitude. NCW represents a powerful set of warfighting concepts and associated military capabilities that allow warfighters to take full advantage of all available information and bring all available assets to bear in a rapid and flexible manner.

Information Sharing is a New Source of Power

NCW translates an Information Advantage into a decisive Warfighting Advantage

Information Advantage - enabled by the robust networking of well-informed geographically dispersed forces

Characterized by:
- Information sharing
- Shared situational awareness
- Knowledge of commander's intent

Warfighting Advantage—exploits behavioral change and new doctrine to enable:
- Self-synchronization
- Speed of command
- Increased combat power

The tenets of NCW are (Alberts, Garstka, & Stein, Network Centric Warfare: Developing and Leveraging Information Superiority, 1999):

- A robustly networked force improves information sharing
- Information sharing enhances the quality of information and shared situational awareness
- Shared situational awareness enables collaboration and self-synchronization, and enhances sustainability and speed of command

- These, in turn, dramatically increase mission effectiveness

Information Age Warfare

Network-centric warfare was followed in 2001 by Understanding Information Age Warfare (UIAW), jointly authored by Alberts, Gartska, Richard Hayes of Evidence Based Research and David S. Signori of RAND (Alberts, Garstka, Hayes, & Signori, 2001). UIAW pushed the implications of the shifts identified by network-centric warfare in order to derive an operational theory of warfare.

Starting with a series of premises on how the environment is sensed, UIAW posits a structure of three domains. The physical domain is where events take place and are perceived by sensors and individuals. Data emerging from the physical domain is transmitted through an information domain.

Data is subsequently received and processed by a cognitive domain where it is assessed and acted upon. The process replicates the "observe, orient, decide, act" loop first described by Col. John Boyd of the USAF (Osinga, 2006).

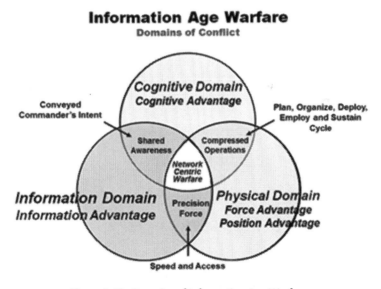

Figure 3-12. Domains of Information Age Warfare

Physical Domain: The physical domain is the traditional domain of warfare. It is domain where strike, protect and maneuver take place across the environments of ground, sea, air and space. Comparatively, the elements of this domain are the easiest to measure, and consequently, combat power has traditionally been measured primarily in this domain. Two important metrics for measuring combat power in this domain, lethality and survivability, have been and continue to be cornerstones of military operations research.

Cognitive Domain: The cognitive domain is the domain of the mind of the warfighter and the supporting populous. This is the domain where battles and wars are won and lost. This is the domain of intangibles: leadership, morale, unit cohesion, level of training and experience, situational awareness, and public opinion. This is the domain where tactics, techniques and procedures reside. The attributes of this domain are extremely difficult to measure, and each sub-domain (each individual mind) is unique. Consequently, explicit treatment of this domain in analytic models of warfare is rare.

Information Domain: The information domain is the domain where information lives. It is the domain where information is created, manipulated and shared. It is the domain that facilitates the communication of information between warfighters. It is the domain where the command and control of modern military forces is exercised, where commander's intent resides. Consequently, it is increasingly the information domain that must be protected and defended to enable a force to generate combat power in the face of offensive actions taken by an adversary. And, in the all-important battle for information superiority, the information domain is ground zero.

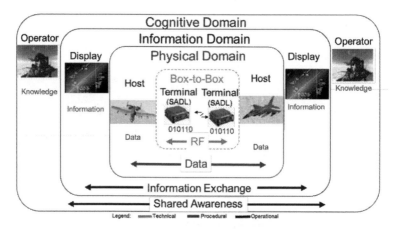

Figure 3-13. Close Air Support Mission: Domain Overlay

Here is a depiction of brain-to-brain collaboration across vast distances enabled by robust physical domains and empowered by consistent information presentations.

Integrating our systems as a network of interrelated capabilities and information is key to meeting the challenges addressed by Gen Jumper in the Global Strike Task Force briefing.

Vonnegut on Professional Competence

"Ninety-eight percent off all people are incompetent at their chosen profession. It is amazing how bad people are at what they do. You can standout as excellent if you are just marginally competent."

Player Piano
By Kurt Vonnegut

Part II. Modeling

In this book, we have divided modeling into four categories: environmental, physical, behavioral, and multi-resolution.

- Environmental modeling represents the terrain, ocean, air space, and weather in which a simulation takes place.
- Physical modeling includes the existence and actions of physical objects and their effects, i.e., detection, target acquisition, and attrition.
- Behavioral modeling is the representation of human and organizational behavior that drives the physical actions.
- Multi-resolution modeling is the mapping of models from one simulation into another.

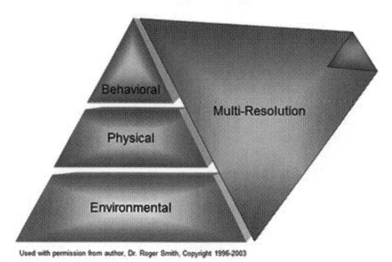

Used with permission from author, Dr. Roger Smith, Copyright 1996-2003

Chapter 4. Environmental Modeling

In legacy models, the environment was frequently not modeled explicitly. Customer and stakeholder requirements are demanding more explicit and higher fidelity environmental modeling. The Close Combat Tactical Trainer (CCTT), in the screen shot below, is an example of these requirements. The components CCTT combine to create a highly complex synthetic battlefield on which soldiers can conduct training in a combined arms environment.

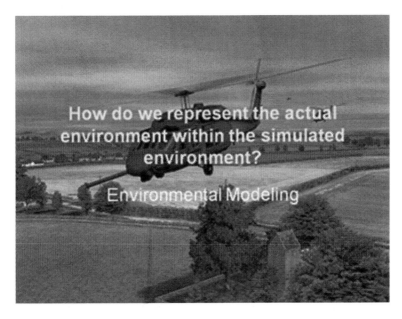

Environmental modeling involves data collection and processing of environmental data, and making choices about dynamic versus static, modeling, level of detail, etc., and often requires tradeoffs.

For example, OneSAF (being integrated into Synthetic Environment Core) has been crafted to be uniquely capable of simulating aspects of the contemporary operating environment and its effects on simulated activities and behaviors. Special attention has been paid to detailed buildings for urban operations including interior rooms, furniture, tunnels and subterranean features, and associated automated behaviors to make use of these attributes.

> **Purposes of Environmental Modeling**
> - Visualization
>   - Backdrop, Pictures, Marketing, Feeling
> - Impedance
>   - Limitations, Visual and Physical Blocking
> - Interactions
>   - Crashing, Slowing, Burning, Dispersion
> - Dynamic Actions
>   - Cloud & Smoke Formation, Tree Movement, Sea State, Weather Fronts
> - Realism & Unpredictability
>   - Weather and Sea Sampling

Environmental details may be necessary for their impact on perception. In this particular situation, it may be important for the environment to hide targets and present an accurate visual scene, but no models for direct vehicle interaction with the terrain are required. This type of high-level terrain visualization was predominant when the customer for virtual simulations was the Air Force fighter community. For example, Tactical Air Combat Maneuver (TACM), geography is represented as a plane surface, with all aircraft positions in x-y coordinates.

Level of Detail

When ground vehicles and helicopters brought ground combat into the virtual world, the requirements for terrain and feature detail increased dramatically. It is essential that surface appearance and trafficability be captured for roads, fields, mountains, riverbeds, and other surface types. Trees must be drawn individually in the foreground and as masses in the background. Buildings are now complex structures that include both external and internal features.

Terrain Level of Detail

Perceived details
- bitmaps over data points
- hills, trees, rivers, rocks

No interaction
- simulated system does not interact directly with terrain details.

Visual detail
- polygon color & lighting
- bit mapped surfaces
- hard surfaces

Modeling detail
- surface trafficability
- foliage density
- tree trunk diameter

The Synthetic Natural Environment (SNE) Conceptual Reference Model (CRM) was created to emphasize the relationships between the contributions of environmental data, models, and other objects. The SNE CRM was developed by Dr. Paul Birkel of Mitre Corporation [1] to describe the issues related to environmental data representation which affect TDB production and consistency within and between systems.

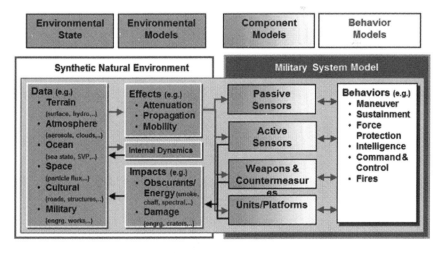

Figure 4-1. Conceptual Reference Model (Birkel, 1999)

Environmental State captures the data about the environment. It describes the environment at any point in time and may be static or dynamic. Environmental Models are constructed to manipulate the environmental state, modify it, and make determinations based upon it.

Component Models are those objects that reside in the environment. They must include algorithms that react to the effects of the environmental state and models. If this is not true, then the environmental data and models will have no impact on the object.

Behavior Models experience secondary effects of the environment. As a results of degraded equipment performance, the behavior model adjusts its plans to compensate.

Data Processing

Computer data processing is any process that a computer program does to enter data and summaries, analyses or otherwise convert data into usable information. The process may be automated and run on a computer. It involves recording, analyzing, sorting, summarizing, calculating, disseminating and storing data. Because data is most useful when well-presented and actually informative, data-processing systems are often referred to as information systems. Nevertheless, the terms are roughly synonymous, performing similar conversions; data-processing systems typically manipulate raw data into information, and likewise information systems typically take raw data as input to produce information as output.

An integrated environmental data base is constructed through a six step process:

- **Collection**. The data must be captured from the real world through some form of surveying process. This mission is increasingly being accomplished by aerial and orbital sensors.
- **Cleaning**. All data sources contain discontinuities that are generated in the collection process. These are eliminated and multiple samplings are stitched together to create a continuous surface.
- **Organizing**. The database is organized such that it can be indexed, archived, and accessed through some standard software tools.
- **Integration**. Multiple data sources are integrated together into a single complex form. This includes attaching vegetation to surfaces, embedding roads in the terrain, and applying surface textures.

- **Transmission**. The integrated database is moved to the host that will use it at simulation run time.
- **Compilation**. The database is compiled into a format that is efficient for simulation execution. This includes cutting the database into sheets that fit the matching page size and converting data formats to those for which the host computer is optimized.

Data Processing (Strickland, ORSA Processes, 2004)

Collection	Integration
• survey the environment (satellite, maps, etc.)	• merge vector, grid, model
• store the results	• generate terrain skin with embedded features and surface data
• vector, grid, and model data	**Transmission**
Cleaning	• move data to the host system
• remove collection process discontinuities	**Compilation**
• synchronize vector and grid data	• create performance-optimized runtime databases
Organizing	
• index and archive	• cut into sheets

Example: The Shuttle Radar Topography Mission (SRTM) is an international research effort that obtained digital elevation models on a near-global scale from 56° S to 60° N,[2] to generate the most complete high-resolution digital topographic database of Earth prior to the release of the ASTER GDEM in 2009. SRTM consisted of a specially modified radar system that flew on board the Space Shuttle Endeavour during the 11-day STS-99 mission in February 2000, based on the older Spaceborne Imaging Radar-C/X-band Synthetic Aperture Radar (SIR-C/X-SAR), previously used on the Shuttle in 1994. To acquire topographic (elevation) data, the SRTM payload was outfitted with two radar antennas (NASA, 2009). One antenna was located in the Shuttle's payload bay, the other – a critical change from the SIR-C/X-SAR, allowing single-pass interferometry – on the end of a 60-meter (200-foot) mast (NASA, 2009) that extended from the payload bay once the Shuttle was in space. The technique employed is known as Interferometric Synthetic Aperture Radar.

The elevation models are arranged into tiles, each covering one-degree of latitude and one degree of longitude, named according to their southwestern corners. It follows that "n45e006" stretches from 45°N 6°E to 46°N 7°E and "s45w006" from 45°S 6°W to 44°S 5°W. The resolution of the cells of the source data is one arc second, but 1" (approx. 30 meter) data have only been released over United States territory; for the rest of the world, only three-arc-second (approx. 90-meter) data are available (Nikolakopoulos, Kamaratakis, & Chrysoulakis, 2006). Each one arc second tile has 3,601 rows, each consisting of 3,601 16 bit bigendian cells. The dimensions of the three-arc-second tiles are 1201 x 1201.

Figure 4-2. Cape Town, South Africa, Perspective View, Landsat Image over SRTM Elevation. Cape Town and the Cape of Good Hope, South Africa, appear in the foreground of this perspective view generated from a Landsat satellite image and elevation data from the Shuttle Radar Topography Mission (SRTM). The city center is located at Table Bay (at the lower left), adjacent to Table Mountain, a 1,086-meter (3,563-foot) tall sandstone and granite natural landmark. The large bay facing right (South) is False Bay. The perspective is computer generated, combining a photograph with elevation data collected using radar. This Landsat and SRTM perspective view uses a 2-times vertical exaggeration to enhance topographic expression. The back edges of the data sets form a false horizon and a false sky was added. Colors of the scene were enhanced by image processing but are the natural color band combination from the Landsat satellite (NASA, 2009).

The elevation models derived from the SRTM data are used in Geographic Information Systems. They can be downloaded freely over the Internet, and their file format (.hgt) is supported by several software developments.

Storing Environmental Data

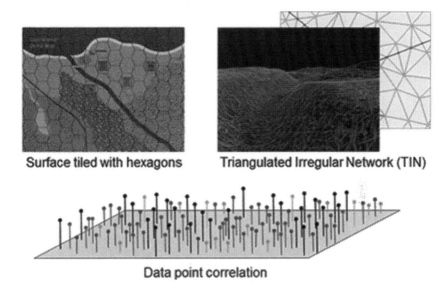

Figure 4-3. Example of methods for storing environmental data

People are often curious about why a hexagon should be chosen as the shape of a tile. There are only three regular geometric shapes that can be used to completely tessellate or tile a space in a constant pattern:

- The triangle
- The rectangle, and
- The hexagon.

All other shapes leave irregular shaped holes between them, or must be arranged in patterns that are irregular and inappropriate for mathematical algorithms.

Triangulated Irregular Networks (TIN) (Tchoukanski, 2005) are a very efficient and flexible method for storing environmental data. These allow the data modeler to represent the environment at different levels of detail and to create smooth transitions between these layers. Triangles, rectangles, and hexagons can form a surface (tessellate a space), but only triangles can be organized and bent to follow any surface. Three points are guaranteed to lie in a definable spatial plane, and triangles can subdivide themselves. The process of TIN'ing a database is essentially one of introducing acceptable levels of error in exchange for storage and processing improvements. Points are added to or removed from the original database to make it represent the surface without redundant information. The process essentially removes data points over flat areas and retains high data point counts for irregular surfaces. This results in an irregular arrangement or triangles across the surface, hence the name of the approach.

Terrain Database Types

The CTDB (Compact Terrain Database) format, an optimized run-time format used by the ModSAF and OneSAF applications. Developed for use in the ModSAF simulation project, the CTDB (Compact Terrain Database) format is widely used but not widely supported in commercial terrain database tools due to its relative complexity. Elevation data in the CTDB can be stored as grids, TINs, or hybrid (grids and TINs). CTDB is composed of (1) terrain skin and bathymetry, (2) Physical features (trees, etc.), (3) abstract features (tree canopies, lakes, soil areas), and (4) road and river networks.

Elevation grid is composed of elevation posts, generally sampled with 125 meter spacing, stored in pages that consist of 32 posts a side. Each elevation post is a 32-bit integer: 19 bits for elevation data, 4 bits for linear features, buildings, trees, and microterrain; 1 bit for alternate diagonalization; the remaining byte is no longer used.

Tin elevation data are stored in patch structures which are generally 500 square meter sections of the CTDB. Terrain elements (TEs) are vertices of polygons that make up the terrain skin. A virtual grid is overlaid on the TEs, and each cell is mapped to the TE that has the largest area within that particular cell.

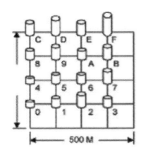 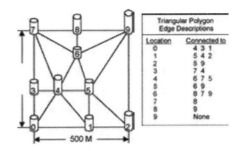

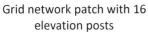

Grid network patch with 16 elevation posts   Triangular irregular network patch

Figure 4-4. Two terrain database types

A patch is a grid that typically has four posts per side, but often has one to five posts per side. For a grid portion of the CTDB, a patch would typically contain 16 elevation posts in a four-by-four grid of regularly spaced posts at 125-meter intervals. Each of its posts is located in the northwest corner of its defined grid square and uses diagonals running from NW to SE (typically) or SW to NE, to create two triangular surface polygons. The top row of posts within the patch would be along the northern border of the patch and the leftmost column of posts would be along the western border of the patch. The posts defining the elevations on the southern and eastern borders of the patch are stored in the adjacent patches on those sides. For grid terrain, data about the features (buildings, roads, trees, micro-terrain, etc.) are stored in the PAT by the patch.

For a TIN portion of the CTDB, up to 25 triangular polygons may be defined within the patch with posts describing the vertices of each triangle such that the polygons completely cover the surface of the patch. Each post has an elevation associated with it. No information on the features are stored with the posts or polygons. For TIN terrain, data about the features are stored for each patch as "terrain elements".

The Polygon Attribute Table (PAT) is a global storage place for sets of attribute value, such as mobility, water content, surface and material categories.

Elevation By Nearest Post

For nearest-post elevation determinations, the elevation is assumed to be identical to the elevation of the nearest post. The surface representation for nearest-post method can be thought of as level squares with sides the same size as the horizontal resolution, with each square centered on the enclosed elevation post, and with each square's elevation value the same as that of the enclosed elevation post. Examples include the Army Airland Battle Environment (ALBE) used in CASTFOREM and the Bresenham algorithm used in Janus 3.x.

The Bresenham LOS algorithm, used in Janus 3.x simulations, is a modification of a fast, efficient algorithm used to determine which pixels light up on a computer screen to depict a straight line. This algorithm uses integer arithmetic for reducing the time for computation. This figure shows an example of selected posts using the Bresenham algorithm (Proctor & Gerber, 2004).

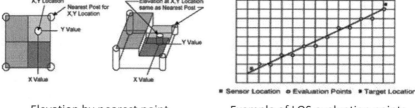

Elevation by nearest point   Example of LOS evaluation points
                                 for the Bresenham algorithm

Figure 4-5. Bresenham algorithm

Elevation by interpolation for grids

In the DYNTACS algorithm, originated in the 1970s and still available through the Combined Arms and Support Task Evaluation Model (CASTFOREM) simulation, each 2D intersection of the line from the sensor to the target and the lines between adjacent elevation posts (both North-South and East-West) are evaluated.

In Janus 4.0+, the line between the sensor and the target is divided into step sizes and evaluated at each step along the line. The Janus 4.0+ approach does not guarantee that all possible elevation obstructions along the LOS are accounted for, but it does provide fewer points for evaluation and therefore a faster calculation, which frees up processor cycles for other activities (Proctor & Gerber, 2004).

Elevation by interpolation for grids

Four-Post Interpolation Procedure

Two two-post interpolations between paired posts on opposite sides of grid square:

$$Z_{ia} = Z_{aa} + (Z_{ba}-Z_{aa})*(X_i-X_a)/(X_b-X_a)$$
$$Z_{ib} = Z_{ab} + (Z_{bb}-Z_{ab})*(X_i-X_a)/(X_b-X_a)$$

Single two-post interpolation between the two previously interpolated points:

$$Z_{ii} = Z_{ia} + (Z_{ib}-Z_{ia})*(Y_i-Y_a)/(Y_b-Y_a)$$

Four-post interpolation within a grid square

Interpolation with TIN

For interpolation, a two-post interpolation is used if the location is on a line between two adjacent posts in grid terrain or on a line between two locations defining the edge of a triangular polygon on TIN terrain. A four-post interpolation is used in grid terrain when the surface is not depicted by triangular polygons and the location lies in the grid square between four elevation posts. Figure 4-6 shows an example of how to perform a four-post interpolation.

In the case of a triangular polygon, representation of the surface, whether the posts are grid or TIN, the elevation of a location is determined by the surface elevation of the enclosing triangular polygon at the x, y coordinate location.

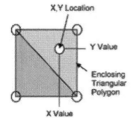
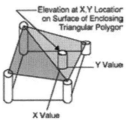

Used in:
- OneSAF Test Bed (OTB)
- Dismounted Infantry Semi-Automated Forces (DISAF)
- Close Combat Tactical Trainer (CCTT).
- ModSAF

Figure 4-6. Elevation by interpolation for triangular polygons

In Janus 4.0+, the line between the sensor and the target is divided into step sizes and evaluated at each step along the line. The Janus 4.0+ approach does not guarantee that all possible elevation obstructions along the LOS are accounted for, but it does provide fewer points for evaluation and therefore a faster calculation, which frees up processor cycles for other activities. For the Janus 4.0+ approach the step size is determined as follows (Proctor & Gerber, 2004):

- First, calculate the x and y differences from the sensor to the target.
- Second, select the largest distance.
- Third, divide that distance by the elevation post spacing.
- Fourth, round to the nearest integer.
- Finally, divide the sensor to target line into that many parts for the step size.

Comparison of Algorithms

Algorithm Name	Post Method	Relative Computation Time	Evaluation Points
Bresenham	Nearest Post	1.00	Fewest
ALBE	Nearest Post	1.33	Very few
Janus	Interpolation	3.09	More
DYNTACS	Interpolation	4.80	Much more

Static Environment

Ground combat simulations have followed the lead of board wargames by tiling the surface with hexagons. Each hex generally contains a single terrain surface type and associated movement and detection modifiers. This approach also encourages the placement of rivers along the edges of hexes and roads between hex centers. Achieving all of these simplifications requires some modifications to the actual data. The level of acceptable variation can drive the size of the hexagons chosen to cover the area.

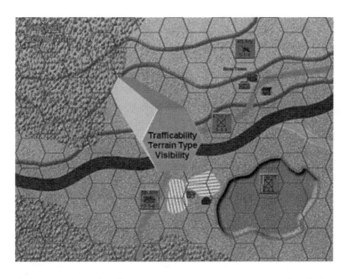

Figure 4-7. Example of a static environment constructive simulation

Constructive simulations like the Corps Battle Simulation (CBS) have settled on a hex size of 3.5 kilometers in diameter (side-to-side, not corner-to-corner). In the Joint Theater Level Simulation (JTLS) the terrain cells are hexagonal with a typical diameter of about 20km.

In the example below (Figures 4-8 and 4-9), researchers from John Hopkins University Applied Physics Lab have characterized the features of the Sea of Japan into a relatively small matrix. In order to model the transmission of sonar waves between a moving target and a sensor array, they simplify the calculations by the matrix categories.

The objective was to characterize all (or a significant portion) of the salient propagation loss behavior of the model, for a specific set of environmental features, in a compact form. These compact abstractions can be used to great effect in synthetic force models, where realistic responses of systems/sensors to the environment is critical.

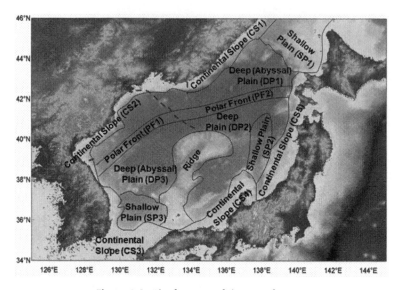

Figure 4-8. The features of the Sea of Japan

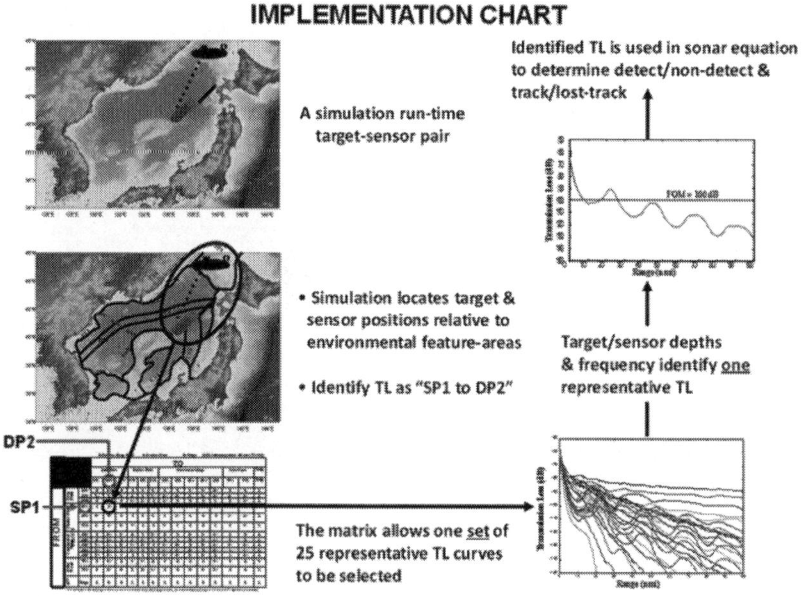

Figure 4-9. Characterization the features of the Sea of Japan into a relatively small matrix

Dynamic Environment

In the real-world the environment is constantly changing. Legacy simulations have largely implemented static environments, but the demands of the customer and the power of computers is raising the demand for dynamic environments. The characteristics described above are very obvious dynamics that will make the virtual world much more realistic.

The Synthetic Theater of War (STOW) program initiated some very aggressive concepts in dynamic terrain. In many cases, the communication of the changes to the environment was simple. It is the visualization of these effects and the actions required to create them that can be very complex. The environmental model itself can be very simple.

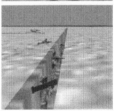

Terrain surface changes
- Anti-tank ditches
- Vehicle survivability positions
- Craters
- Infantry trenches
- Berms

Other obstacles
- Concertina wire
- Dragon's teeth
- Minefields

Damage to vertical structures

Figure 4-10. Examples of dynamic terrain in STOW

In the real world the environment is constantly changing. Legacy simulations have largely implemented static environments, but the demands of the customer and the power of computers is raising the demand for dynamic environments. The characteristics described above are very obvious dynamics that will make the virtual world much more realistic.

Independent
- Weather movement: clouds, rain, wind
- Sea state: storms, daily tide
- Daylight: sunrise, sunset, dark
- Smoke & Dust: clouds, raising, dispersing

Interaction
- Holes: artillery craters, engineering artifacts
- Tank treads: tracks, destruction
- Terrain morphing: engineering, construction
- Feature modification: building damage, trees burned

Figure 4-11. Characteristics and examples of dynamic terrain features

For example, building rubbleing, smoke, and artillery dust is implemented in Janus v. 7.1. In CASTFOREM, a variable unit resolution down to the individual weapon system level, resolution of terrain is also variable. Battlefield environments modeled include static weather, dynamic obscurants, such as smoke and dust, nuclear and chemical contaminants, and electronic warfare.

Classic Problems in Interpretation

Terrain Elevation Points is a classic example of misinterpretation. Given four points in a grid, it is clear how they should be joined around the outside edge. However, there is not enough information in the points themselves to indicate how they should be connected across the diagonal. Is this picture supposed to form a small valley or a small ridge? Different simulation systems may interpret this same data in two different ways.

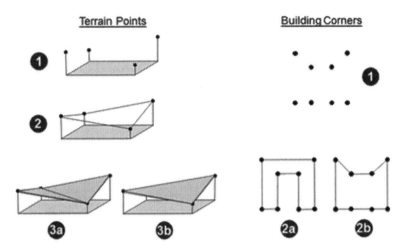

Figure 4-12. On the left, terrain points as shown in (1) and connected in (2) can be interpreted as either (3) or (4). Without additional information, the interpretation is uncertain. Building corners also pose a problem. The points defined in (1) on the right can be connected in two different ways as shown in (2a) and (2b).

Building Corner Points is another classic example. Given the corner points of this building, it is not clear what the shape of the building is. If you know the order in which they are connected, then it may form a U-shaped building with a south-facing inner courtyard. If you connect them in a clockwise order, then the building is shaped like a crown with a courtyard intruding from the north side.

Environmental Standardization

As independent simulation systems are tied together through DIS and HLA, we create a virtual environment in which each simulation has its own representation of the environmental data. This data has come from different sources, been processed into unique formats, and is displayed with unique tools. As a result, the consistency of environmental data between two simulation systems varies widely. Some find their data sets very closely correlated, others find huge discrepancies in their environments.

Figure 4-13. All four screenshots are different representations of the MOUT site at Fort Benning, GA. Each has a different level of detail and uses more or less data than th3 others, probably in different formats

Therefore, the military has been searching for some means to create equivalent environmental representations for all simulation systems. This must be accomplished in an environment in which vendors use proprietary data and formats.

The Synthetic Environment Data Representation and Interchange Specification (SEDRIS) is a project aimed at providing a standardized approach to storing and exchanging environmental data. SEDRIS is...

- A method (a language) for unambiguously describing the environment (both real and simulated)
    - Semantics and relationships of data elements
    - All environmental domains
    - Expressed in a data representation model
- Also a mechanism to share and interchange the described environment
    - Sharing and re-use
    - Ease of access and software development (API)
    - Tools and applications
- An infrastructure technology for expressing and sharing environmental data

Chapter 5. Physical Modeling

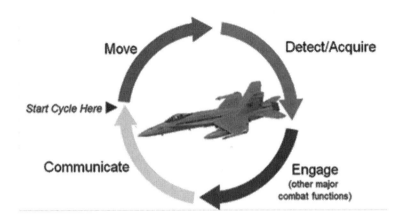

Typical military simulations have focused on the actions of movement, sensor detection, and engagement/attrition. Additional models were viewed as supporting features for "the big three." As simulations became more object oriented, each object gained a level of independence that required significant communication with other objects to operate. These communications could be included as part of the infrastructure, or they could be modeled according to the phenomena that allow the operations in the real world.

Object specialization also means that the "engagement" actions may no longer be the universal activity. Instead, the operation phase will execute the primary operation for which each specific object exists in the simulation.

This change in perspective still considers objects as vehicles or creatures that dynamically interact with the environment. However, buildings and similar cultural features may also be treated as independent objects. These features are currently treated as part of the environment.

Chapter 6. Movement Modeling

Movement is phenomenology that customers and stakeholders are requiring higher resolution for simulations and simulators. Physics-based movement, automated and topology smart movement meet many of the requirements. However, constructive simulation may require less fidelity and may incorporate movement points, automatic route planning, and terrain feature movement.

Grid Hopping

Most people are familiar with binary engagement/attrition models that use grid hopping. These are used in most entertainment board games. The piece that is attacking (making a move) wins in all cases in games like Chess and Checkers. In Stratego and other wargames, the strength of the units involved are considered, but the decision is the same – you live or die.

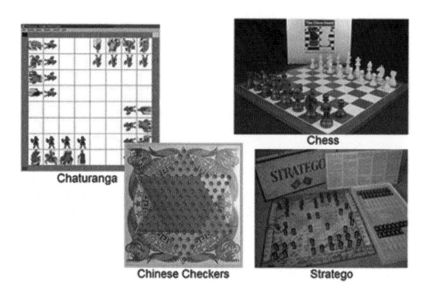

Figure 6-1. Example of board games that use grid hopping

Sector (Pistons) Movement

The picture below shows a once-standard piston model representation of a theater with contiguous sectors. In such a case (e.g., the old NATO Central Region), the simple treatment requires the assumption that both sides fall back as necessary to avoid overexposing flanks. The battles in the various sectors are then assumed to be independent frontal attacks characterized by close combat between opposing ground forces. Thus, the battles described involve classic close-combat attrition warfare without the complications of air forces and long-range fires. Moreover, because we use scalar strengths, we ignore unit structure and the interactions of various weapons types.

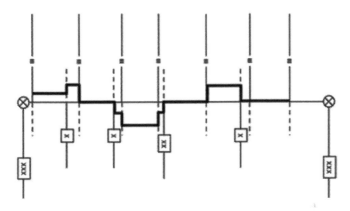

Figure 6-2. Depiction of how a piston model works

Movement-Points Movement

In many cases, the surface has been tiled into some regular shape. Each surface tile then contains information about the mobility effects it has on an object. The hexagon maps used by wargames are an excellent example of this. Each hexagon represents a different surface type with a unique impact on the mobility of the vehicle. Each vehicle has a calculated number of movement-points that it must spend to cross the terrain.

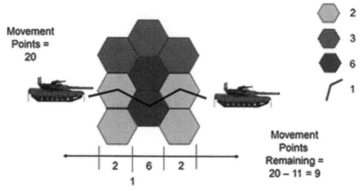

Figure 6-3. Movement-points example

In this situation, the movement of the vehicle must be handled in segments. Each segment overlays a specific surface tile or hex. The sum of the times across a set of tiles determines the amount of time required to cover a specific distance.

Under these conditions, it is also common to assign movement effects to the edges of the tiles. The edges are convenient places to insert the effect of features like rivers and minefields.

It is customary for roads to run from hex-center to hex-center to insure that they are not co-linear with rivers. Instead, intersect them at hex edges, thus indicating the presence of a bridge.

Surface Tiled with Hexagons

You have probably seen games that use a hexagon grid to regulate the playing area before, but its invention was a very special solution to a particularly vexing problem. You see, prior to the hexagon grid, strategy games were always designed around a square grid, as you see on a chessboard. However, western military schools that were adopting military simulation gaming to their curriculum (going back to the Prussian army of the 1860s) found a fatal flaw in the square grid system. That is, a unit moving one space diagonally along that grid, moves 45% further than a unit moving one space horizontally or vertically along it. Thus, it was difficult to simulate realistic rates-of-movement using a square grid without putting bizarre restrictions on, or doing all kinds of mathematic calculations, to cover this anomaly of diagonal movement.

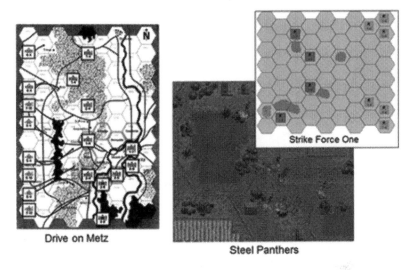

Figure 6-4. Example of wargames with surfaces tiles as hexagons

In the 1950s, the RAND Corporation, a Think Tank that helped solved these kinds of problems, found the solution to this square grid problem in the honeycomb of bees. The hex grid was developed to simulate a circle of options for every unit located on it. That is, the unit could move to any of six adjacent spaces and each would cover and equidistant amount; this was the end of the square grid diagonal movement dilemma. Suddenly, realism in game grids took on a hexagonal look that is still used to this day.

Bald Earth Movement

Bald earth movement is the most fundamental movement assumes a bald surface and moves the object in a straight line. Determining the ending point is as simple as rate*time = distance, and then adding the X and Y component of the distance to the current location of the vehicle.

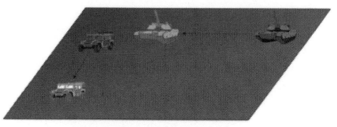

Set heading, speed, start time

Rate*Time = Distance

20 km/hr * 30 min = 10 km

Figure 6-5. Depiction of bald earth movement

Terrain & Feature Movement

Adding detail to terrain will require additional knowledge about the capabilities of the vehicles. This may be done by assigning nominal movement rates under each terrain type, or by creating a movement model that replicates the physics of movement. The latter approach assumes that the detail needed to feed the physics equations is resident in both the terrain database and the vehicle description data. The JCATS entity level wargame uses the equation shown to determine that actual speed of an object. Each factor is derived from the environmental conditions, unit state, weather, and daylight within the simulation.

Set Objective: Position or Vector
Terrain & Features Modify Instantaneous Heading & Speed

Speed = min(order_speed, max_speed*tafficability*slope_factor)* weather_factor*lighting_factor*fatigue_factor*supression_factor

Figure 6-6. Terrain and feature modeling in JCATS

Physics-based Movement

This flowchart shows the movement model created for the automated forces within the Close Combat Tactical Trainer (CCTT). The CCTT ground-vehicle mobility model is based on a general first-principle dynamics model (physics-based movement). The algorithms used in this project were drawn from the validated NATO Reference Mobility Model. That model provides detailed movement capabilities for different vehicles operating in different modes across different terrain types.

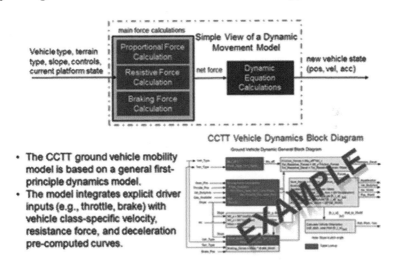

Figure 6-7. Physic-based movement in CCTT

- Proportional Force Calculation (calculates the raw force generated by the platform)
- Resistive Force Calculation (calculates the resistance force felt by the platform due to friction and such)
- Braking Force Calculation (calculates the force generated by the platform's brakes)
- Dynamic Equation Calculations (maps the net force to movement in the form of distance, velocity, acceleration)

Automatic Route Planning

Automatic route planning allows the analyst to focus on higher issues such as the overall scheme of maneuver. It reduces the intrusion of the analyst into C2, especially synchronization of the forces if C2 is modeled explicitly to allow mission-driven planning.

- CONCEPT: provide an algorithm by which units can automatically find their own routes.
  - allows the analyst to focus on higher issues such as the overall scheme of maneuver
  - reduces the intrusion of the analyst into C2
  - units can still be given explicit routes if desired, or closely grouped intermediate objectives
- ALGORITHMS: based on graph theory
  - could be a satisfying algorithm (not guaranteed to find an optimal route)
  - might be an optimal algorithm
  - "optimal" may mean fastest, or shortest, or safest, etc.
- EXAMPLES
  - A* search, Johnson's algorithm, Dijkstra's algorithm, hill climbing
- Techniques:
  - Based on geophysics
  - Based on finite difference schemes
- Examples:
  - VR-Forces – MäK Technologies
  - UMBRA Simulation Framework – Sandia Labs

Topology Smart

Movement model selects path from topological map while maintaining the objective or waypoint for which it is heading. The route traveled is a function of the topology. Topology smart has been applied to networks in NCW simulations. Topology smart is a function of:

- The number of polygons
- Constraint curves
- Tradeoff (objective) curves
- Refresh rate
- And rest solution

Figure 6-8. Example of topology smart movement

When human look at a map they use many visual clues to help them quickly determine the shortest (or nearly so) path between their position and their Goal. However, a computer does not have the luxury of visual data processing. A computer must produce similar results by considering each segment of a potential path and comparing it to what has been previously calculated. Eagle Corp-level combat model uses this method.

A* Search Algorithm

One of the most powerful algorithms for performing shortest path calculations is the A* Search algorithm. This technique explores all of the options available from the current position. The cost of movement along each segment is added to an estimate of the cost of achieving the goal from that new position. The goal estimate is usually the minimum cost that is possible. For example, if the cost were solely dependent upon the distance, then the estimate would be the straight line distance from the point being considered to the goal.

- CONCEPT: Provide an algorithm by which units can automatically find their own routes.
    o This allows the analyst to focus on higher issues such as the overall scheme of maneuver
    o This lessens the intrusion of the analyst into command and control, especially synchronization of the forces if C2 is modeled explicitly to allow mission-driven planning

- o Units can still be given explicit routes if desired, or closely grouped intermediate objectives
- Algorithms: based on graph theory
  - o Could be a satisficing algorithm (not guaranteed to find an optimal route)
  - o Might be an optimal algorithm
  - o "Optimal" may mean fastest, or shortest, or safest, etc.

However, we should note that the algorithm is based on "cost" values. This means that the selection of the shortest path may include information about distance, terrain features, congestion, vegetation, speed limit, and anything else that can be quantified into an impact on movement.

Example: In this example, we want to move from S, the starting point, to G, the Goal.

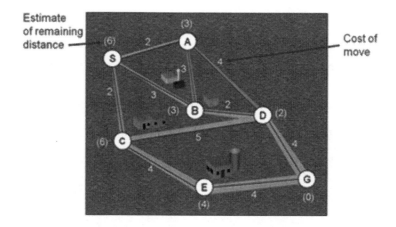

Figure 6-9. Setup for in the A* search algorithm

The first segments to consider are between S and the points A, B, and C. Here we take the cost of moving on the segment from S to A, 2, and add it to the estimate of the remaining distance to the goal, 3, to obtain a cost of 5. The process is repeated for segments S-B and S-C. See Figure 6-10.

Start with an examination of the route from S to A which has cost 2.

Then examine of the route from S to B which has cost 3, etc.

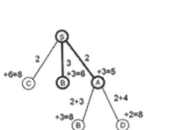
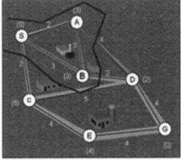

Figure 6-10. First step in the A* search

The same process is repeated starting with A and moving toward the Goal segment by segment (paths S-A-B and S-A-D).

The estimated remaining distance from both A is (3), to both B and D so examine both paths.

Figure 6-11. Second step in the A* search

The same process is repeated starting with B and moving toward the Goal segment by segment (paths S-B-A and S-B-D). See Figure 6-12.

Figure 6-12. Third step in the A* search

Both paths A and B have led to a common path, D. In this case, we continue with the process, starting with D on the B-path (since it has minimum cost) and moving toward the goal segment by segment (paths S-B-D-C and S-B-D-G).

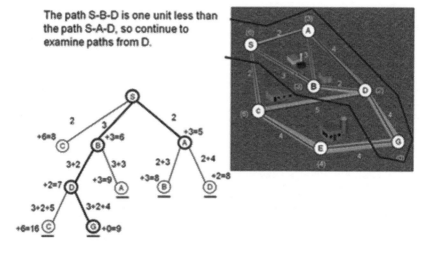

Figure 6-13. Fourth step in the A* search

At this point we can go no further along path S-B-D-G (since it has reached the Goal). Thus, since the path S-B-D-C has not reached the goal, we compare the path S-C-E and calculate its cost, which we compare to the path S-B-D-G. In this instance, path S-C-E has a high cost, so there is no need to continue assessing path S-C-E-G. Therefore the optimum path produced by the A* Search is path S-B-D-G. See Figure 6-14.

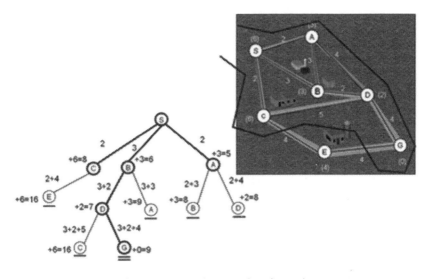

Figure 6-14. Final step in the A* search

Grid Registration

To improve performance, a simulation should always contain a grid system across which ground and sea surface vehicles are moving (2D, or through which aircraft and submarines move, 3D). This grid system is like an advanced version of the checkerboard. Each time a vehicle moves to a new location, it should be registered into the grid it occupies. This grid registration is often managed in the form of a linked-list of vehicle that is attached to each grid cell.

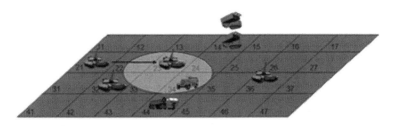

- Grid 14 → V71, V109, V1212, V10101
- Vehicles registered into geographic grid during movement
- Improve LOS, sensor, and interaction performance

Figure 6-15. Example of grid registration

This information can be used to quickly determine the congestion in a specific area. Such information can improve the performance of any algorithm that needs to locate other vehicles near it (sensor scans, movement congestion, target selection, etc.). The grids can change an $O(N^2)$ problem to an $O(N)$ problem. For example, 1000 objects doing range checks on each other requires 1,000,000 calls to the range equations. Grids can enable each object to identify the grids under its sensor fan, then go straight to the list of objects in registered grids. If the fan covers 8 grid cells that average 2 objects each, then range calculations are reduced to $1000 * 8 * 2 = 16,000$ (on average). In some cases you may even eliminate range checks entirely and just perform LOS or other detection algorithms on the vehicles. The grid check may also replace the range calculations as a tool for pre-selecting potential targets.

Beyond 2-D Movement

We use a three-dimensional (3-DOF) coordinate system to describe the motions of an aircraft in three-dimensional space. The center of the coordinate system is found at the centroid of the aircraft - roughly between the wings in the middle of the fuselage. A centroid is an imaginary point around which all rotation takes place. The three axes intersect at the centroid at right (90 degree) angles to each other. Movement along or around one axis does not necessarily involve any movement on or around the other two.

- 3 Dimensional—aircraft rotation axes
    - yaw - vertical axis rotation
    - roll - longitudinal axis rotation
    - pitch - lateral axis rotation
- 3-D Mathematics
    - Euler angles
    - axis angle
    - rotation matrices
    - quaternions
- Other degrees of freedom:
    - 3+3 DOF
    - 5-DOF
    - 6-DOF

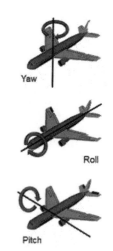

Figure 6-16. Depict of 3-DOF: Yaw, Pitch and Roll

One of the most common types of transformations that are simulated in 3D Engines is rotation. There are a number of ways of representing rotations: in matrices, quaternions, angle vectors, and in degrees and radians. The most accurate and least limited way of storing them is in matrices, called rotation matrices or direction cosine matrices. A matrix is a mathematical concept that is basically a grid of numbers. When these numbers are applied to another matrix, or number, or most commonly in a 3D engine, a point, in the correct order, they can modify the values of what they are being applied to.

3+3 DoF

- 6 coordinates define the geometric state of an object
- Same as 3-DoF plus (θ, φ, ψ) for pitch, yaw & roll about the centroid
- Orientation state generated by perturbing desired (ideal) orientation with an orientation error model and moderating the transition from one orientation to another without a damping response model (i.e., instantaneous response in orientation)

6-DoF

- Same as Pseudo 3 + 3 DoF except that the orientation is generated by solving $\partial L_i / d\partial$ = Torque, for the orientation coordinates
- Requires knowledge of moments of inertia with respect to principal body axes (distribution of mass in body)

Behavioral—agent Based Movement

From a modeling and simulation perspective, one novel way to perform movement (or C2) is to employ a polyagent modeling construct to produce emergent planning behavior. A polyagent is a combination of a persistent agent (an avatar) supported by a swarm of transient agents (ghosts) that assist the avatar in generating and assessing alternative (probabilistic) futures.

The ghosts in the model employ pheromone fields to signal, identify, and act on threats and opportunities relative to the goals, which are then reported back to the avatars for integration and decision-making. Pheromones can be thought of as markers to tell units whether an area is attractive or unattractive for future exploration.

$$F_n = \frac{\theta \cdot RTarget_n + \gamma \cdot GTarget_n + \beta}{(\rho \cdot GNest_n + \beta)(Dist_n + \varphi)^{\zeta + \alpha(RThreat_n + 1)} + \beta}$$

- Behavioral evolution and extrapolation
- Each avatar generates (a) a stream of ghosts samples the personality space of its entity.
- They evolve (b, c) against the entity's recent observed behavior.
- The fittest ghosts run into the future (d),
- and the avatar analyzes their behavior (e) to generate predictions.

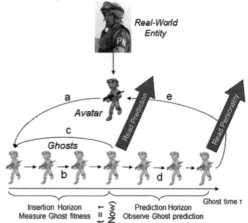

Figure 6-17. Agent-based moment algorithm

An avatar is a computer user's representation of himself/herself or alter ego, whether in the form of a three-dimensional model used in computer games, a two-dimensional icon (picture) used on Internet forums and other communities, or a text construct found on early systems such as multi user domains (MUDs). It is an "object" representing the embodiment of the user.

F_n is the resultant attractive force exerted by neighbor n and $Dist$ is the distance to the target if it is known.

Chapter 7. Detection Modeling

Detection modeling, like movement modeling, is requiring higher fidelity models that are explicitly executed in simulations. However, detection modeling is more complex than outlined here. For example, a radar detection model for radar cross section (RCS) requires signatures for the objecting being detected. Thus, signatures must be modeled as well. This section will present only the fundamental aspects of detection modeling.

Perfect Detection

The most primitive form of detection is perfect detection. Models may choose to allow all vehicles to know the location and status of all other vehicles. This eliminates the computational complexity and storage requirements for locating, identifying, storing, and distributing perceived data about enemy and friendly vehicles (Smith R., Military Simulation Techniques & Technology, 2006).

- Every object knows the true location of every other object.
- There are no models of sensors or processors.

Figure 7-1. Perfect detection in board games

Even simulations that choose more detailed detection equations often choose to allow perfect detection among all objects that belong to the same side of combat – the same country, alliance, force structure, etc.

Grid Probability Areas

The next level of detection modeling is proximity detection. Vehicle in a bounding area or tile, like a hexagon, are usually allowed 100% detection of other vehicles in the same bounding area. Outside of that area, modelers often implement one or two different options. They may allow no detection outside of the bounding area, or they may allow detection within tiles that are adjacent to the tile containing the sensor. In the latter case, a probability of detection is assigned to each terrain type. This number is compared with a randomly generated number to determine whether detection of a specific vehicle has occurred (Smith R., Military Simulation Techniques & Technology, 2006).

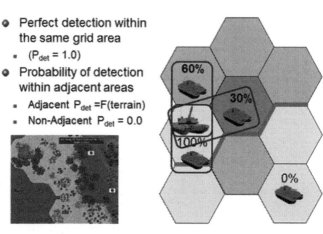

- Perfect detection within the same grid area
  - (P_{det} = 1.0)
- Probability of detection within adjacent areas
  - Adjacent P_{det} = F(terrain)
  - Non-Adjacent P_{det} = 0.0

Figure 7-2. Grid probability area detection example

Detection Range

A circular detection range around the object with the sensor is a very obvious model of the sensor. Even when the sensor is modeled in much greater detail, these types of range circles are used as a first step to filter out targets that could not possibly be detected (or that model designers have decided not to detect – such as friendly vehicles) (Smith R., Military Simulation Techniques & Technology, 2006).

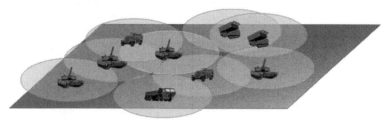

- Complete circle—no field of view/field of regard
- Terrain line-of-sight (LOS) is separate

Figure 7-3. Detection range example

3D Detection

The next step is to define detection areas based on the capabilities of the specific sensor. This begins to differentiate the capabilities of one vehicle or sensor package from another.

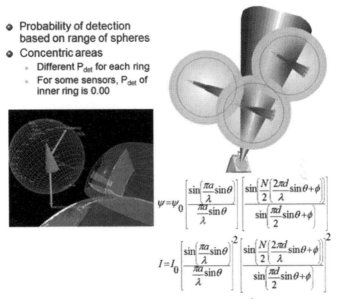

- Probability of detection based on range of spheres
- Concentric areas
  - Different P_{det} for each ring
  - For some sensors, P_{det} of inner ring is 0.00

Figure 7-4. 3D detection area example

Probability areas assign one or more circles or spheres around the vehicle. Each ring is assigned a probability of detection. This can be a single-value that applies to all vehicles that enter the area, or it can be expressed as a matrix in which the number varies according to characteristics of the target. With a little more work the P_d values can be converted into a function that models a constantly increasing P_d as the target moves closer to the sensor. Ψ is diffraction with fringe effect and I is the intensity of the wave (Strickland, Mathematical and Heuristic Models of Combat, 2009).

Target Acquisition

Target acquisition is generally concerned with the detection of points of interest (POIs), and their subsequent recognition and identification. John Johnson originally quantified these criteria in the 1950s. He investigated the relationship between the ability of the observer(s) to resolve bar targets (one black bar and one white bar equate to one cycle) through an imaging device and their ability to perform the tasks of detection, recognition, and identification of military vehicles through the same optical sensor (Volimerhausen, Jacobs, & Driggers, 2003).

Johnson observed basic spatial resolution dependence for target acquisition tasks. Table 1 shows the original Johnson criteria for several targets and three discrimination tasks, as noted below (Johnson, 1958).

1. **Detection** - "There's something out there"
2. **Recognition** - "It's a tank"
3. **Identification** - It's a T72 tank".

From these various visual tasks, and for nine different targets, generally accepted values were determined. Over the years since the Johnson criteria were developed, some minor modifications were made; but the general information has remained unchanged. The changes to the visual tasks are: the detection task has changed from one cycle to 0.75 cycles, the recognition task was broken into optimistic at 3 cycles and conservative at 4 cycles; and identification is generally accepted to be 6 cycles.

Table 1. Johnson Criteria - The number of just resolvable cycles required across a target's critical dimension for various discrimination tasks.

Target (Broadside)	Resolution per Minimum Dimension (Line Pairs)		
	Detection	Recognition	Identification
Truck	0.90	4.50	8.00
M-48 Tank	0.70	3.50	7.00
Stalin Tank	0.75	3.30	6.00
Centurion Tank	0.75	3.50	6.00
Half-Track	1.00	4.00	5.00
Jeep	1.20	4.50	5.50
Command Car	1.20	4.30	5.50
Soldier (Standing)	1.50	3.80	8.00
105 Howitzer	1.00	4.80	6.00
Average	1.00±0.25	4.00±0.35	6.40±1.50

Search is the process of seeking a target. There are two simple types of searches: random and scanned, with or without detection. Two of the more complex search are glimpse and continuous searches. Detection rates can be constant or can be varied with range. Detection rates can also vary based on the size and orientation of the target to the sensor. They can also be probabilistic

Dynamic Tactical Simulation (DYNTACS) was developed in the 1960's, but is still used in legacy models such as CASTFOREM. It considers line-of-sight as a regression model to fit field data. Input variables are range, enemy height, enemy speed, terrain complexity, and probability of looking in the sector containing the enemy.

JCATS Detection Algorithm

The algorithms in Joint Conflicts And Tactical Simulation (JCATS) were derived from the Night Vision Electro-Optical Lab (NVEOL) model. At the start of the simulation a 128128 matrix is generated from the NVEOL Detection Map used in JANUS(A) 5.0. The Detection Map consists of one hundred values representing a log normal distribution. JCATS randomly selects from the Detection Map while filling a 128 × 128 matrix. All viewer/entity pairs in the simulation are then hash mapped to the matrix. This means that for a given simulation run, a given viewer/entity pair will always have the same acquisition threshold value. However, due to the random fill of the matrix, the same viewer/entity pair may (and probably will) have a different threshold in subsequent runs.

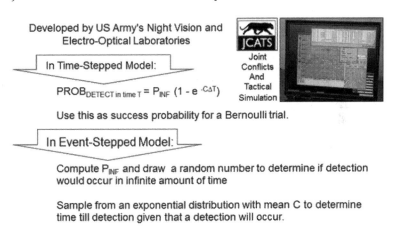

Figure 7-5. Night Vision Electro-Optical Lab (NVEOL) model

NVEOL models both optical sensors and thermal sensors.

The algorithm requires computation of P_{INF} = probability to detect target given infinite amount of time as a function of N. Experimental data has validated values for N_{50} = number of cycles to achieve 50% success in detection (approximately 6.4 cycles for optical and thermal sights). P_{INF} can be modeled as Normal(N_{50}, $0.6\, N_{50}$)

Sensor Characteristics

To add another layer of detail, detection can be achieved through the specific pairing of the sensor and the target. In this type of model the sensor will operate within a defined maximum range and over a specific search area. A target that falls into this area is subject to detection at different levels based on its type or size and the profile it presents to the sensor. The detection will be further adjudicated by geometry factors like range to the target and the off-set angle to the sensor. Finally, terrain and weather data may be used to determine whether the sensor has a clear line-of-sight to the target and the degree to which visibility is degraded by weather conditions.

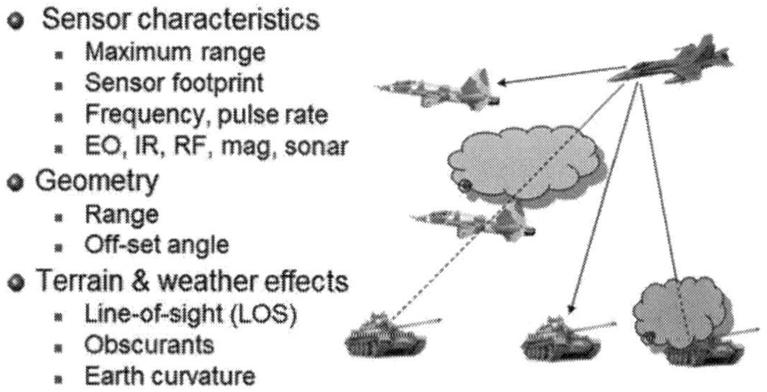

- Sensor characteristics
  - Maximum range
  - Sensor footprint
  - Frequency, pulse rate
  - EO, IR, RF, mag, sonar
- Geometry
  - Range
  - Off-set angle
- Terrain & weather effects
  - Line-of-sight (LOS)
  - Obscurants
  - Earth curvature

Figure 7-6. Example of sensor characteristics, geometry, terrain and weather effects

Target Characteristics

Other factors that affect detection are target characteristics. These include vehicle camouflage, color and pattern of the paint on a vehicle, radar cross section, its IR signature, movement, cavitation, magnetic mass, obscurant on the battlefield, and the earth's curvature.

- Camouflage
- Color & Pattern
- Radar Cross Section
- IR Signature
- Movement
- Cavitation
- Magnetic Mass
- Obscurants
- Earth Curvature

Figure 7-7. Examples of target characteristics

The Basic Target Acquisition Process

The target acquisition process is usually a probabilistic event and has two conditions:

1. Physical preconditions must exist

    (a) Line-of-sight (Observer-Target)

    (b) Within detection range

    (c) Observer is looking toward target, sensors turned on, etc...

2. Actual detection must occur

In the basic target acquisition process, two conditions must be present. First, a physical precondition must exist that would allow target acquisition, such as Line-of-Sight (LOS), within detection range, sensors on, etc. Second, an actual detection must occur. This is usually a stochastic process.

Levels of Target Acquisition

Cueing is a prerequisite of target acquisition. It occurs when an observer is alerted to possible presence and/or approximate location of a target (e.g., by a muzzle flash, a sound, movement, etc.). the levels of target acquisition are:

- **Level 1** is Detection: Observer sees object and decides that it has military interest (e.g., distinguishes a vehicle from some natural terrain feature)

- **Level 2** is Classification: Observer discriminates among finer classes of targets (e.g., tank, vs. infantry fighting vehicle)
- **Level 3** is Recognition: Observer recognizes actual target type (e.g., M1 vs. T80 tanks)
- **Level 4** is Identification Friend or Foe (IFF): Observer decides whether the identified target is friend or foe.

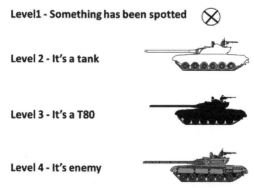

Figure 7-8. Levels of target acquisition

Basic Target Acquisition

This is a simplified version of the target acquisition process, where the sensor is on (making glimpses) and a decision is made about target detection. This decision is based on probabilities. If the target is found the process stops. Otherwise, the process loops back through the glimpse phase until a target has been detected. In more sophisticated models, as time progresses the chance of target detection improves.

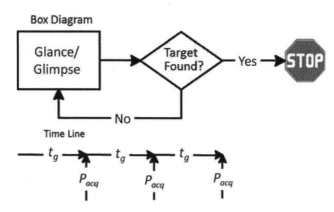

Figure 7-9. Flow diagram of basic target acquisition

Detailed Target Acquisition

In this more detailed model, the point sensor looks at a scene, in its area of detection, processes inputs from objects in its detection area, assesses the data it collects, and finally decides if a target is present. This process should allow the sensor to make distinctions between object of interest, like other vehicles, and terrain and its features, like rocks and trees.

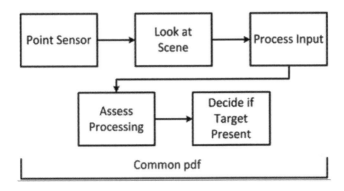

Figure 7-10. Detailed target acquisition flow model

Probability of Line-of-Sight

In order for a combat simulation to model modern combat accurately, it must incorporate methods to evaluate whether or not one entity detects another entity on the battlefield. While detection is certainly a simple concept to understand, it is not always a simple concept to model. Detection is dependent upon numerous factors; one of these factors is line-of-sight (LOS).

The LOS factor is simply a yes or no answer: does one entity have LOS to another entity? Various algorithms exist for computing LOS in open terrain; the most widely used open terrain LOS algorithms are discussed in (Champion, 1995) and (Champion, 2002). While these algorithms have subtle differences, almost all open terrain LOS algorithms use a common basis to determine whether a sensor has LOS to a target: comparison of slopes. First, using given elevation data, the slope of the line connecting the sensor to the target is computed. Next, for sufficiently many intermediate points on the ground directly between the sensor and the target, the slope of the line connecting the sensor to the intermediate point is computed (this process of computing "intermediate" slopes is usually accomplished by successively "stepping" along the underlying terrain skin from the sensor location to the target location). If any of these intermediate slopes exceeds the slope from the sensor to the target, then the sensor does not have LOS to the target. Figure 7-11 shows a simplified illustration of LOS determination in open terrain.

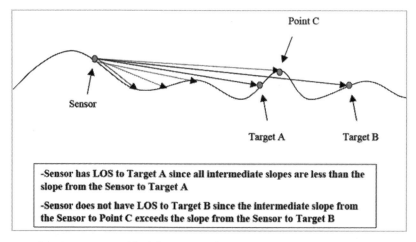

-Sensor has LOS to Target A since all intermediate slopes are less than the slope from the Sensor to Target A

-Sensor does not have LOS to Target B since the intermediate slope from the Sensor to Point C exceeds the slope from the Sensor to Target B

Figure 7-11. A simplified illustration of LOS determination in open terrain

Computing LOS in urban terrain is even more complicated, as man-made structures, such as buildings, can block LOS. Additionally, targets might be located inside fabricated structures, effectively blocking LOS from any location. Thus, an algorithm used to compute LOS in urban terrain must evaluate whether the underlying terrain skin blocks LOS—and, at a minimum, fabricated structure. Algorithms that are more sophisticated would consider vegetation, obscurants, etc.). Figure 7-12 demonstrates the complexities introduced to LOS by fabricated structures in urban terrain.

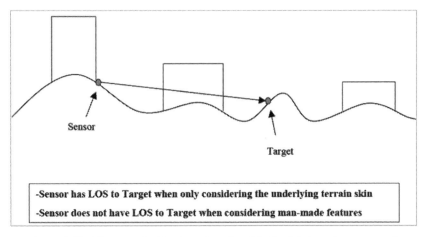

Figure 7-12. The complexities introduced to LOS by fabricated structures in urban terrain.

Probability of Line-of-sight

As one can imagine, LOS algorithms can become quite complex in combat simulations. Combine this with the fact that, by nature, LOS calculations must be frequently executed, and one can understand why LOS algorithms can consume a large amount of computing time in a combat simulation. As a result, researchers have focused a large amount of effort to find more efficient or equally useful methods for computing LOS. One of these approaches is to use probability of line-of-sight (PLOS) in place of traditional LOS calculations.

PLOS calculations use digital elevation data to estimate the probability that LOS exists at a given sensor-to-target range (this is ground range, not slant range; see Figure 7-13), sensor elevation, and terrain type. A combat simulation using a PLOS methodology would calculate PLOS estimates for each sensor-to-target range, sensor elevation, and terrain type in pre-processing (prior to the actual simulation run) and store them in a PLOS "look-up table." Then, when the simulation needs to make an LOS determination during the simulation run, it simply uses the given sensor-to-target range, sensor elevation, and terrain type to determine the probability that LOS exists using the PLOS look-up table. Subsequently, a simple uniform random number draw and comparison to the PLOS will determine whether LOS exists.

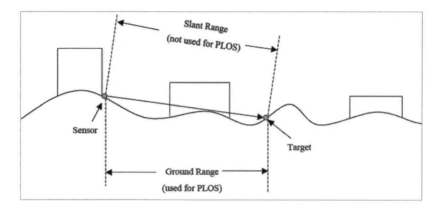

Figure 7-13. PLOS calculations use digital elevation data to estimate the probability that LOS exists at a given sensor-to-target range.

Thus, PLOS requires the same data as traditional LOS algorithms, but it is less accurate (a traditional LOS algorithm is deterministic, while computing LOS using PLOS estimates is stochastic). The reason, then, for even considering the use of PLOS is that traditional LOS algorithms necessitate the use of computationally expensive computations during the simulation, increasing run time. PLOS calculations, in contrast, require the use of computationally expensive calculations during pre-processing and computationally inexpensive table look-ups during the simulation. Thus, if there is an ample pre-processing time budget for a simulation, the use of a PLOS methodology can greatly reduce simulation run time.

PLOS is clearly not appropriate for simulations involving few entities and few events or interactions—traditional LOS algorithms will suffice. However, when the number of entities and interactions becomes large, simulation run time can become unacceptably long. In these situations, it may be worthwhile to sacrifice LOS accuracy in order to reduce simulation run time. It is in exactly these situations that a PLOS methodology can be useful.

PLOS Curves

A PLOS curve is simply a plot of PLOS versus sensor-to-target range. All other variables, such as sensor elevation, target elevation, and terrain type, are fixed for a given PLOS curve. Thus, PLOS curves consolidate PLOS estimates in graphical form and provide a visual picture of how PLOS behaves as a function of range. An example of a PLOS curve is shown in Figure 7-14. In general, PLOS curves will be monotonic decreasing—we would expect PLOS to decrease as range increases. The shape of PLOS curves, however, will vary. PLOS curves may be linear, piecewise linear, concave, convex, or a combination of these shapes (as shown in Figure 7-14).

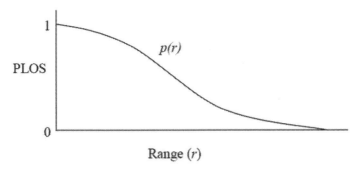

Figure 7-14. PLOS curve

Joint Warfare System Use of PLOS

The Joint Warfare System (JWARS) is a campaign-level model of military operations that has been developed for use by the Office of the Secretary of Defense, the Joint Staff, the Services, and the Warfighting Commands (Stone, 2001). Because of a willingness to sacrifice LOS accuracy in order to improve simulation run time, JWARS currently employs a PLOS methodology for combat scenarios in open terrain. In JWARS, PLOS is a function of three parameters: sensor-to-target range, sensor elevation, and effective terrain roughness. The third parameter, effective terrain roughness, is a function of sensor-to-target range and the aggregate qualities of the underlying terrain; more detailed information on the calculation of effective terrain roughness can be found in (Blacksten, 2002) and (Blacksten, 2004).

The PLOS estimates computed by JWARS are approximated by fitting a functional approximation to the PLOS data (using the sensor-to-target range, sensor elevation, and effective terrain roughness as parameters) rather than using a PLOS lookup table. Consequently, when JWARS needs to perform a LOS calculation, it does not explicitly calculate LOS based on a comparison of slopes. Instead, it simply evaluates a function in order to determine the probability that LOS exists – the PLOS. Then, as previously described, a uniform random number draw and comparison to the PLOS will stochastically determine whether LOS exists.

As demonstrated by JWARS, the use of PLOS in combat simulations that model open terrain is not a novel idea; however, at the time of this research, a PLOS methodology had yet to be applied to urban terrain.

Line-of-sight Algorithms

An LOS algorithm determines whether the terrain at any point blocks any of the one or more rays that are projected from the sensor to the target.

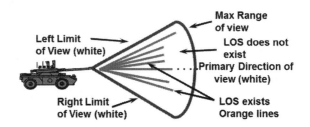

Figure 7-15. Line-of-sight concepts

LOS blockage is determined by stepping through various locations along the three-dimensional ray between the sensor and the target and by comparing the elevation of the ray with the terrain elevation at those locations. If the terrain is higher than the ray at any location, that ray is considered blocked and no more steps along the ray are taken.

Alternatively, slopes from the sensor to the terrain at evaluated points can be compared with the slope from the sensor to the target to determine blockage.

Line-of-sight Models

An LOS algorithm determines whether the terrain at any point blocks any of the one or more rays that are projected from the sensor to the target.

LOS blockage is determined by stepping through various locations along the three-dimensional ray between the sensor and the target and by comparing the elevation of the ray with the terrain elevation at those locations. If the terrain is higher than the ray at any location, that ray is considered blocked and no more steps along the ray are taken.

Simulation models often choose to model LOS in one of two ways. LOS is explicitly modeled in models that store terrain representation data, such as grids and surfaces (i.e., TINs). On the other hand, when models do not contain this data, the model may store the expected results of LOS and uses a lookup table when LOS results are required. This may be accomplished, for example, by probability of LOS.

Explicit Grid Line-of-sight Model

In a simple explicit LOS model, grid data between sensor and target is analyzed to determine if there is a higher elevation grid between sensor and target. This can be accomplished sequentially from grid to grid with the If-Then-Else statement. In this example, grid 5 would impede LOS if it followed the elevation indicated by the dashed elevation extension.

```
Explicit grid line-of-sight model:
    For j = 2,8
        IF (Slope1,j>= Slope1,8) THEN No Line-of-sight
        Where Slopei,j = (Elevj + Vegj - Elevi) / Distancei,j, and
VegTARGET = 0
Explicit Surface Line-of-Sight Model:
Implicit Probability of LOS:
    PREPROCESSING: To determine PLOS
    EXECUTION: To determine whether to allow LOS between two
entities
Implicit Intervisibility Segment Length:
    PREPROCESSING: To determine Average Intervisibility
    EXECUTION: To determine whether to allow LOS between two
entities
```

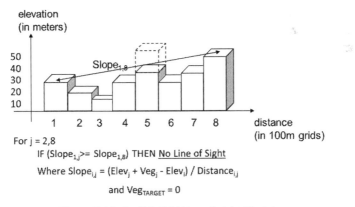

Figure 7-16. Explicit Grid Line-of-sight Model

Explicit Surface Line-of-sight Model

When surface data is available, a cross-section from sensor to target is examined in a similar manner as with grids. Implicit Probability of LOS: Implicit LOS models are the type of models described earlier with hexagon terrain models. In the simplest case, a probability LOS is assigned for vehicle in adjacent hexagons, based on terrain type. These models do not contain terrain elevation data (they are essentially flat even though we may describe a hex as being a mountain), and sensors do not really "see" targets in the model.

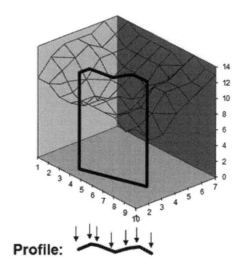

Figure 7-17. Explicit Surface Line-of-sight Model

Implicit LOS models are the type of models described earlier with hexagon terrain models. In the simplest case, a probability LOS is assigned for vehicle in adjacent hexagons, based on terrain type. These models do not contain terrain elevation data (they are essentially flat even though we may describe a hex as being a mountain), and sensors do not really "see" targets in the model. A slightly more sophisticated approach is described below.

- PREPROCESSING: To determine PLOS
  - STEP 1: For each terrain area (e.g., 5km by 5km square area) draw random pairs of locations and count the number of times there is LOS between them (using a high-resolution algorithm).
  - STEP 2: The proportion of times there was LOS becomes the probability of LOS between any two entities in the area.
- EXECUTION: To determine whether to allow LOS between two entities
  - If they are in the same terrain area, use PLOS as recorded
  - If they are in different areas, use average PLOS for each area between them, inclusive. (Other rules possible)

Implicit Intervisibility Segment Length

Another approach skirts around LOS and determines average intervisibilty range in an area, by averaging the intervisibilty ranges of evenly spaced rays in the detection area. This method is demonstrated in Figure 7-18. To determine whether LOS is allowed between a sensor and target, the separation range is compared to the calculated average intervisibilty range.

- PREPROCESSING: To determine Average Intervisibility,
  - STEP 1: For each terrain area (e.g., 5km by 5km square area), select random locations. Then for each location, for each of a set of evenly-spaced headings, determine the range to the furthest visible point.
  - STEP 2: Average the ranges to determine the average intervisibilty range in the area.
- EXECUTION: To determine whether to allow LOS between two entities,
  - Determine their separation and compare with the average Intervisibility range ("cookie-cutter" method).

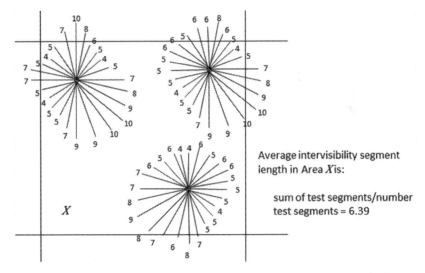

Figure 7-18. Implicit Intervisibility Segment Length: Another approach skirts around LOS and determines average intervisibilty range in an area, by averaging the intervisibilty ranges of evenly spaced rays in the detection area. To determine whether LOS is allowed between a sensor and target, the separation range is compared to the calculated average intervisibilty range.

Intermittent LOS Model

To add a layer of complexity, Intermittent LOS can be modeled as a Markov chain model, with two or more states in the state space. A transition probability matrix describes the probability of moving from one state to another state. In the example above, if the target is initially invisible to the sensor, in the next time step it may still be invisible with probability $P(I|I)$, or it may change state to Visible and Not Acquired with probability $P(VNA|I)$, or it may transition to state Visible and Acquired with probability $P(VA|I)$.

Transition Rates

Φ_η = Rate of Gaining LOS = 1/(Mean Time Target Invisible)

Φ_μ = Rate of Losing LOS = 1/(Mean Time Target Visible)

Φ_λ = Rate of Acquiring Target = 1/(Mean Time to Acquire)

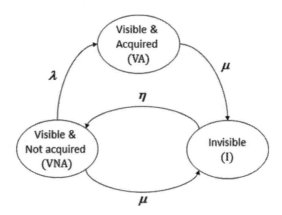

Figure 7-19. Markov chain model of target acquisition

Two Different Engagement Outcomes Possible

- Target Killed
- LOS Lost (and Target Not Killed)

Probability That Target Killed Before LOS Lost

$$P_{K(LOS)_{xy}} = \frac{\alpha}{\alpha + \mu_{XY}}$$

The Greek lower-case letters λ, μ, and η, described the rates of gaining LOS (λ and η), or losing LOS (μ).

We now develop a mathematical model for probabilities that target is in the each of the three states shown in the figure above

- Probability that Target Is Invisible, $P_I(t)$
- Probability that Target Is Visible, But Not Acquired, $P_{VNA}(t)$
- Probability that Target Is Visible and Acquired (and Hence Available to Be Engaged), $P_{VA}(t)$

The model is given by a system of three differential equations involving transition rates.

Database Support for Intermittent LOS Model

The US Army has developed methodology for generating database consisting of numerical values for the rates of gaining LOS η, losing LOS μ, and acquiring targets λ (required input for VIC model). Extensive databases exist for various specific pieces of terrain (from NIMA).

The Glimpse Model

Two basic stochastic detection model paradigms were developed in the OEG56 report (Koopman, 1946). These are known as the ***glimpse model*** and the ***continuous looking model***. Most subsequent detection models have included aspects of one or the other of Koopman's basic models, so we begin our study of target acquisition modeling by reviewing the foundations. In this section we examine the glimpse model; the next section will deal with the continuous looking model. The underlying assumption of the glimpse model concept is an observer has intermittent chances to detect a target ("glimpses").

Time-Stepped Model

Suppose that an observer has intermittent chances to detect a target. For example, an observer who is scanning along a wood line in search of enemies will scan past the location of a target once per pass. Rotating scan radar will have its beam on the target once during each rotation. We call each such intermittent opportunity to detect, a glimpse. We index successive glimpses $i = 1, 2 \ldots$ starting when the search begins. Let

g_i = the probability of detection of the target on the ith glimpse, (assuming that the search is still continuing, i.e. that the previous $i-1$ glimpses have failed to detect the target, and assuming that a target is present.)

Each glimpse can be considered to be a Bernoulli trial with success probability g_i. Simple probability argument, assuming independence of the repeated Bernoulli trials, yield

$$\text{Prob(no detect on first } n-1 \text{ glimpses)} = \prod_{i=1}^{n-1}(1-g_i)$$

$$\text{Prob}(first \text{ detect on first } n^{th} \text{ glimpse}) = g_n \cdot \prod_{i=1}^{n-1}(1-g_i)$$

$$\text{Prob(detect } \varepsilon \text{ first } n \text{ glimpses)} = \prod_{i=1}^{n}(1-g_i)$$

An important special case is where all the glimpse probabilities are equal, $g_i = g_j = g$. Then the above results become

$$\text{Prob(no detect on first } n-1 \text{ glimpses)} = (1-g)^{n-1}$$

$$\text{Prob}(first \text{ detect on first } n^{th} \text{ glimpse}) = g(1-g)^{n-1}$$

$$\text{Prob(detect } \varepsilon \text{ first } n \text{ glimpses)} = 1-(1-g)^n$$

In this case the number of glimpses until the first detection is distributed according to the geometric probability distribution.

Normally the time step for a fixed time step simulation would not be the same as the glimpse interval for a particular sensor, especially in a combined arms scenario where many different sensors are in use. Suppose that the time step interval is significantly longer than the glimpse interval, say

$$TIME\_STEP = n * GLIMPSE\_INTERVAL$$

for an integer n. Then detection within a time step corresponds to detection with n glimpses, so for each time step and for each target we would compute either

$$\text{Prob(detect } \varepsilon \text{ first } n \text{ glimpses)} = 1 - (1 - g)^n$$

If n is small enough that the observation conditions are approximately constant within a time step, or

$$\text{Prob(detect } \varepsilon \text{ first } n \text{ glimpses)} = 1 - \prod_{i=1}^{n}(1 - g_i)$$

if conditions change.

Estimating the Numeric Value of g

For the geometric distribution, the expected value of the number of glimpses required to attain detection is

$$E(n) = 1/g$$

which is a useful relationship for computing numerical estimates of g based on experimental data.

Suppose we perform a number of experiments in which observers view successive glimpses of a target scene until they detect the target. Let x be the number of glimpses required in the jth run of the experiment. Then the sample mean, $\bar{X}$, is a maximum likelihood estimate of $E(n)$, and we can solve for the estimated glimpse probability

$$g = 1/\bar{X}.$$

If we are interested in the behavior of g under different observational conditions, we could set up a series of experiments for each observational condition and table the resulting set of estimates.

Event-Stepped Model

The concept here is to draw a random variable $N \sim \text{geometric}(g)$, with the acquisition time

$$TIME_{acq} = N \times \text{GLIMPSE\_INTERVAL}.$$

How many time-steps does it take for an observer to acquire a target, assuming a time-stepped, stochastic combat model using the glimpse detection model? Assume that detections are independent from one time-step to the next, i.e., what happened on an earlier time-step is irrelevant to later detections. The following apply:

- Length of a glimpse = 10 seconds; Time-step = 1 minute
- Probability of detection in one glimpse, $g = 1$
- Use these numbers as $U(0,1)$ random numbers (RN): .9295, .6137, .3867, .6001, ...

Number of glimpses per time step:

- n = timestep_interval/glimpse_interval = $60/10 = 6$

$Prob$[detect in each time-step] = $1 - (1 - g)^6 = 1 - .96 = 1 - .5314$
$= .4684$

For each timestep, draw X from $U(0,1)$. If $X \leq .4686$, a detection occurs that time-step. Therefore, it takes three time-steps to acquire, since .3867 is the first RN less than .4686.

Intermittent Glimpses

So far, we have considered simple searches within a specified search space. We will now examine target search from the perspective of the number of glimpses made over time, and the duration of those individual glimpses, rather than the area of the search space. The first type of search to detect a particular target is the intermittent search, in which we make a succession of brief glimpses, as shown in Figure 7-20, where a surface-to-air missile (SAM) site is tracking a penetrating aircraft (Strickland, Mathematical Modeling of Warfare and Combat Phenomenon, 2011).

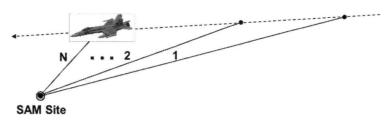

Figure 7-20. A surface-to-air missile (SAM) site is tracking a penetrating aircraft

The most important parameter in this process is the glimpse probability, which gives the probability of seeing the target on the ith glimpse. This probability will actually vary from glimpse to glimpse depending on changes in some physical conditions such as range, speed, illumination, weather, etc. We will assume here that the value of is known, and will not discuss the methods of actually estimating it.

Knowing these values, we can determine the effectiveness of the target acquisition and tracking system in terms of the probabilities of target detection and the expected number of glimpses required to detect the target.

For a particular search, we number each successive glimpse from the beginning of the search. For the ith glimpse, the probability represents the probability of detecting the target provided the target has not been detected previously. We will also assume that the target is seen for the first time after N glimpses so that. Using these assumptions and the definition of compound probabilities, we define the probability that detection fails to occur after the first n glimpse as, so that

$$Q(n) = Q(\text{detection fails to occur on each of the first n glimpses})$$

The probability of detecting the target in the first n glimpse is found from

$$P(n) + Q(n) = 1$$

which represents the basic condition for probabilities; i.e., the sum of the probability of an event occurring and the probability of the same event not occurring is equal to one. Hence, the probability of detecting the target on the first n glimpses is given by

$$P(n) = 1 - \prod_i (1 - g_i)$$

We may now obtain the probability of first detection of the target on the nth limpse from

$$P(n) = Q(\text{detection fails to occur on each of the first } n-1 \text{ glimpses})$$
$$\times P(\text{detection on the } nth \text{ glimpse})$$

It is interesting to note here that the relationship between the probability of detecting the target in the first n glimpses and the probabilities of first detecting the target on the ith glimpse is given by $P(n) = p(1) + p(2) + \cdots + p(n)$.

Also, $P(N) = \sum_n p(n) = 1$.

We may find the expected number of glimpses required for target detection by computing the expected value of N given above.

The main objective when searching for a target is to make the probability of detection as large as possible. In many cases, detection in the fewest number of glimpses may be more important in order to conceal the searching beams. Consequently, we may use both $P(n)$ and $E(n)$ as measures of effectiveness (MOE) of a target acquisition system, depending on the particular objective.

Continuous Looking Model

In looking for a target with a sensor (e.g., radar) operating continuously, information is also being received continuously. For such cases, the probability that detection occurs at a given instant is zero. The continuous looking model is the second of Koopman's basic detection model paradigms. Most modern high resolution combat simulations use a continuous detection model both because it is more flexible than the glimpse model and because it fits easily into the structure of an event scheduled simulation (Strickland, Mathematical Modeling of Warfare and Combat Phenomenon, 2011).

The continuous looking model is based on a detection rate function, D (t), which has the property that the probability of detecting a target in a short time interval is proportional to the length, Δ, of the interval and is given by:

$$\text{Prob(detect in } [t, t + \Delta T]) = D(t) * \Delta T$$

For initial simplicity, assume for now that $D(t) = D =$ constant for all t.

In a longer time interval of length $T = N * \Delta T$, we can use the mathematics of the glimpse model to yield:

$$\text{Prob(detect in length } T) = 1- \text{Prob (fail } N \text{ times)}$$
$$= 1 - (1 - D * \Delta T)^N$$
$$= 1 - (1 - D * T/N)^N$$

In the limit as N approaches infinity and ΔT approaches zero, with the product $T = N * \Delta T$ held constant, this is equivalent to

$$P(\text{detection occurs in time } t) = 1 - e^{-Dt}$$

This last expression is recognized as the cumulative distribution function (CDF) of the (negative) exponential distribution most frequently used to model the time required to detect a target. The exponential distribution is the continuous analogue to the discrete geometric distribution, which appeared in the glimpse model. In fact, the mathematics of the two basic models is essentially equivalent. Both of the basic models make the assumption of independence of successive increments of time and thus both have the memory less property. That is, "The fact that you have been searching for five minutes without success does not influence the probability of achieving a detection in the next ten seconds. It is the same as if you had just begun to search now."

The exponential distribution is a high variance distribution. It possesses maximum randomness. Short detection times are the most likely, but arbitrarily long times can occur. The expected value of the time to detect given by the exponential distribution is analogous to that of the geometric,

$$E(T) = 1/D$$

The probability of detection after time t has elapsed is a meaningful measure of effectiveness (MOE) of the system. Another useful measure is the mean time to detection MTTD, which represents the expected value of time t to detect the target. The probability of detection with MTTD or rate of detection is an exponentially distributed random variable, and the CDF gives the probability of detection after time t has elapsed:

DYNTACS Curve Fit Model

Dynamic Tactical Simulation (DYNTACS) was developed in the 1960's, but is still used in legacy models such as CASTFOREM. It considers line-of-sight as a regression model to fit field data. Input variables are range, enemy height, enemy speed, terrain complexity, and probability of looking in the sector containing the enemy.

DYNTACS LOS uses a method similar to JANUS. The DYNTACS terrain representation is also a square grid of elevation posts. An identical four-point linear interpolation scheme is used to determine sensor and target elevation. Instead of breaking the sensor-target line into equal parts, DYNTACS LOS determines every point at which the sensor-target line crosses a facet edge in the square lattice. The algorithm then determines these elevations by linear interpolation on the two known elevation posts. These interpolations are deferred in code until their values are needed. DYNTACS steps through these grid crossing points from sensor to target and compares the sensor to target slope with the sensor to intermediate point slope. If the slope from sensor to intermediate point ever exceeds the sensor-target slope, the algorithm breaks out of the loop and returns false. If the algorithm steps through to the target it returns true.

NVEOL Acquisition Algorithm

The algorithms in JCATS were derived from the Night Vision Electro-Optical Lab (NVEOL) model. At the start of the simulation a 128 × 128 matrix is generated from the NVEOL Detection Map used in JANUS(A) 5.0. The Detection Map consists of one hundred values representing a log normal distribution. JCATS randomly selects from the Detection Map while filling a 128 × 128 matrix (Strickland, Mathematical and Heuristic Models of Combat, 2009). All viewer/entity pairs in the simulation are then hash mapped to the matrix. This means that for a given simulation run, a given viewer/entity pair will always have the same acquisition threshold value. However, due to the random fill of the matrix, the same viewer/entity pair may (and probably will) have a different threshold in subsequent runs.

NVEOL models both optical sensors and thermal sensors. An example with a thermal sensor is:

1. Determine target signature S_T (e.g., heat difference from background)

2. Compute attenuated target signature at sensor location as function of atmospheric conditions:

$$SR = S_T \, e^{-sR} \text{ for range } R.$$

3. Convert the received signal S_R into a "spatial frequency", based on the resolution of the sensor; use a "minimum resolvable temperature", or MRT curve to translate S_R into F_S.

4. Compute number of system resolution cycles the sensor can use on the target given target height and range

$$N = H \, F_S \, / \, R$$

5. Compute P_{INF} = probability to detect target given infinite amount of time as a function of N. Experimental data has validated values for N_{50} = number of cycles to achieve 50% success in detection (approximately 6.4 cycles for optical and thermal sights). P_{INF} can be modeled as Normal($N_{50}, 0.6\,N_{50}$)

6. Adjust P_{INF} to account for sensor scanning:

$$PROB_{DETECT\ in\ time\ t} = P_{INF}\ SCAN(t)$$

$$= P_{INF}\,(1 - e^{-CT})$$

where C is the search rate with respect to FOV and Field of Search.

Using NVEOL

The following is an example target acquisition using continuous-looking, NVEOL model.

How many time-steps does it take an observer to acquire a target, assuming a time-stepped, stochastic combat model using the continuous-looking detection model? Assume that detections are independent from one time-step to the next, i.e., what happened on an earlier time-step is irrelevant to later detections. The following apply (Strickland, Mathematical Modeling of Warfare and Combat Phenomenon, 2011):

- Probability of detecting the target given an infinite amount of time = 0.85
- Detection function $D(t) = 0.05$ det/sec;   Time-step = 1 minute
- Use these numbers as U(0,1) random numbers: .6001, .5528, ...

$$P_{INF} = 0.85$$

$$D(t) = 0.05$$

$$Prob[\text{detect in each timestep}] = P_{INF}(1 - e^{-D \cdot T})$$

$$= 0.85\,(1 - e^{-.05 \times 60})$$

$$= 0.85\,(1 - 0.0498)$$

$$= 0.8077$$

- Each time-step, draw X from $U(0,1)$. If $X \leq 0.8077$, a detection occurs.
- Therefore, it takes <u>one time-step</u> to acquire, since, $0.6007 < 0.8077$.

Chapter 8. Communications Modeling

The basic components of a C4ISR model include terrain databases, propagation models, and communications and network models for both voice and data traffic, and sensor models. Communications modeling in many legacy models assumed perfect communications, which is not reality. Customer and stakeholder requirements are demanding more explicit, higher fidelity models of communications.

Communications Model Effects

Communications modeling adds the effects associated with information exchange on the battlefield. In the past, many assumed that information is exchanged magically, instantly, and perfectly. This is far from an accurate representation of the battlefield.

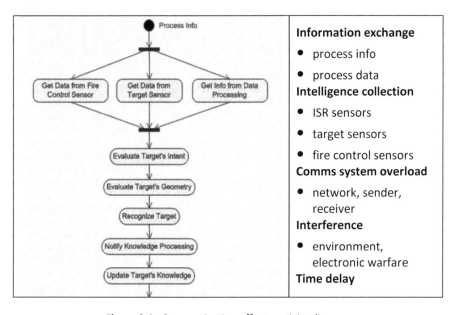

Figure 8-1. Communication effects activity diagram

Communications modeling injects intelligence collection, system overload, interference, time delays, etc. As our forces become more and more "digitized", the impacts of information exchange become a bigger part of the combat experience. Command and computer data exchange have also become the major focus of training exercises. The Netcentric C4ISR of the

US Army Future Combat Systems adds additional complexity to communications modeling.

Perfect Communications

The simplest communications modeling usually involves the creation and passing of messages "perfectly". There is no representation of communication networks or the effects of the environment on wave propagation. These models allow explicit command and control and often use the command hierarchy as a pseudo-network for limiting communications between units. It may be possible for communications in all forms to reach any unit, regardless of its destination, situation, etc.

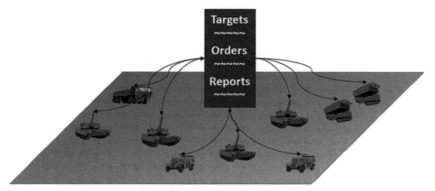

- Shared information, no representation of comms
- Software-to-software message delivery

Figure 8-2. Example of perfect communications

Direct Message Passing

Modeling communications has always been of secondary interest in most simulations. However, with advances of the digital age, the military has become as much an information organization as it is a combat organization.

Adding communications effects to existing models can be achieved at several different levels. The most basic is to apply abstract communications. Models in this category are easily "strapped on" to existing simulations. An abstract model of communications does not represent communications networks or the messages that would traverse them. Instead it reflects the impacts of good and poor communications.

This can be done by consulting the command hierarchy of the unit performing an operation.

In this example, an artillery battery is conducting a fire mission. The accuracy of its fires on enemy locations is dependent upon target information received through communications networks. The artillery unit may consult the health or status of its parent unit. When that unit's health declines, the artillery unit assumes that targeting information would be partially or totally unavailable. In response, the unit would intentionally add error to its firing locations.

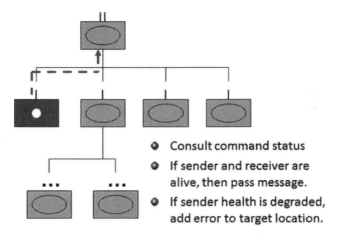

Figure 8-3. Direct message passing example

Broadcast Messages

Communications signals can be transmitted onto the battlefield. Each receiver then must determine whether the signal is accessible from his position and state. The range to the transmitter is one major factor, followed by terrain degradation, earth curvature, traffic contention, and internal state of the receiver. At some point, received signals may further be subjected to a probabilistic assignment of the quality of the receipt. This final step is the most basic representation of the physics of transmission.

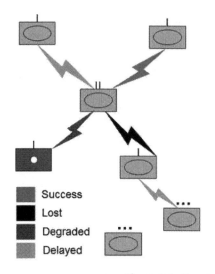

Figure 8-4. Broadcast messages example

Communication Network Models

Mathematical Structure of Networks

The term "network" has become a ubiquitous synonym for any connected system; other synonyms like "grid," "chain" or "mesh" are likewise creeping into operational language. Very few who use these words exhibit understanding that the terms have very specific technical definitions. Although a detailed review of the mathematics of networks is included as an appendix, it is worth discussing an example here. A "grid," for example, is technically a "lattice of degree four," which means there are exactly four links connected to each node. This means there are no shortcuts in a grid and it has a very rigid structure, properties, among others, that make it a poor candidate for a real operational network. The popularity of this term as a descriptor for operational military networks suggests that it is used metaphorically rather than technically, but the fact that it is not even a good metaphor seems to be missed altogether.

This is not an arcane distinction—one should care about the specific mathematical properties of a network for two very practical reasons. The first reason is that different networks have dramatically different properties, and blindly choosing a network type simply because it is popular is the worst kind of engineering, resulting in systems with the wrong properties for the tasks they are required to perform. Second, many of the characteristics which concept developers ascribe to new operational concepts, such as "adaptation," "self-synchronization," "networked effects" or "robustness," have specific mathematical definitions that can be derived from the science of networks. Any model of distributed, networked combat that ignores the mathematics of networks would therefore inappropriately represent combat in the Information Age.

Basic Combat Network Structure

Distributed, networked warfare should be represented as a "combat network." The resulting Information Age combat model should have the mathematical structure of a network, which at its most basic level is represented by nodes connected with links (Cares, 2004).

For the purpose of an Information Age combat model, nodes are defined as elements in a process that are sensors, deciders, influencers, or targets. By definition, sensors receive observable phenomena from other nodes and send them to deciders. Deciders receive information from sensors and make decisions about the present and future arrangement of other nodes. Influencers receive direction from deciders and interact with other nodes to affect the state of those nodes. A target is a node that has some military value but is not a sensor, decider or influencer. In addition, all nodes have a characteristic called "side" (e.g., blue, red; friend, foe, neutral; etc.).

Nodes can be "contracted" so that the functions or values of more than one node can be contained in a smaller number of nodes. For example, a single node can contain the attributes of a sensor, influencer, decider or a target. This allows for representation of a decider and sensor on the same platform.

Contracting a group of sensors, deciders, influencers, and targets into a group of nodes (with one sensor, decider, influencer and target each) leads to an interesting result that approximates Lanchester equations. This shows the potential for a very helpful result of the network model approach: if traditional models can be represented using this framework, then traditional warfare and distributed, networked warfare can be compared using the same model. This comparison is currently impossible with existing models.

Nodes are linked to each other by directional connections called "links." An example of a link is an observable phenomenon that emanates from a node and is detected by a sensor is a link. In this case, links might be radio frequency (RF) energy, infrared signals, reflected light, communications or acoustic energy. Phenomena detected by sensors are communicated to deciders, constituting another kind of link. Deciders issue orders to nodes and influencers interact with other nodes, typically in an effort to destroy or render useless those nodes. These are further examples of links. Note that links are not necessarily information technology (IT) connections between nodes, but represent something more functional—the operative interactions between nodes.

Combat Networks

The links and nodes as defined above constitute a combat network. Figure 8-1 graphically represents the most basic combat network, while Figure 8-2 represents a two-sided system. The use of different line styles in these figures underscores that the links are not homogenous. For clarity, these styles will be omitted in later figures. In addition, the nodes on each side are represented by different colors.

The relationships in the networks in Figure 8-5 and 8-6 have the following characteristics:

- Sensor logic does not equate to decision-making capability
- All sensor information that passes to an influencer must do so through a decider; "sensor-to-shooter" is allowed, "sensor-to-bullet" is not
- Targets can be vehicle platforms without sensing, influencing or decision making capability
- Targets, influencers and sensors are located by sensors; there is no direct path from targets, influencers or disconnected sensors directly to deciders

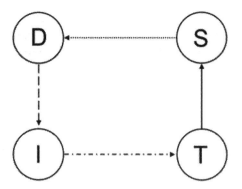

Figure 8-5. Simplest Combat Network

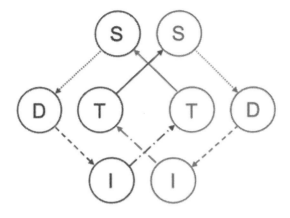

Figure 8-6. Simplest Two-Sided Combat Network

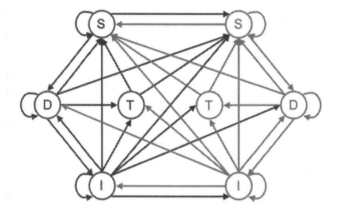

Figure 8-7. Simplest Complete Combat Network

Figure 8-7 represents the simplest complete combat network that is created from these assumptions. This diagram represents all the ways that sensors, deciders, influencers and targets interact meaningfully with each other.

Dimensions and Complexity

The two dimensional surface of this paper obscures the inherent complexity of this "simple" network: there are at least 36 different dimensions in which this network operates. This dimensionality is more evident in a different type of network representation, the adjacency matrix. The adjacency matrix in Figure 8-8 is an equivalent representation of the network in Figure 8-7. A "1" in the matrix indicates that there is a link from node listed at the head of the row to the node listed at the head of the column. A "0" indicates that there is no link between those nodes. Note that the connections are directional from rows to columns. For example, the blue side "*I*" has a link from its own side's decider, *D*, and the red side influencer, *I*, but not from its own side's sensor, S, or target, *T*, or from the red sensor, *S*, decision node, *D*, or target, *T*. Counting all the matrix entries filled with a "1" provides the dimensionality of the simplest, complete combat network, 36. Recall that this is the simplest complete model; one could include many more targets, sensors, decision nodes, and influencers.

	S	D	I	T	S	D	I	T
S	1	1	0	0	1	0	0	0
D	1	1	1	1	1	0	0	0
I	1	1	1	1	1	1	1	1
T	1	0	0	0	1	0	0	0
S	1	0	0	0	1	1	0	0
D	1	0	0	0	1	1	1	1
I	1	1	1	1	1	1	1	1
T	1	0	0	0	1	0	0	0

row maps directionally to column = 1, 0 otherwise

Figure 8-8. Adjacency Matrix

Not only is this structure high dimensional, but it is also complex, in the sense that there is an extraordinary large number of different sub-networks that can be created from this combat network. In general, the number of different sub-networks that can be created from an $N \times N$ matrix is $2^{(N^2)}$. This number gets very large even for small values of N. Figure 5 is a plot of $2^{(N^2)}$.

Values of N larger than 17 can create more possible sub-networks then there are particles of matter in the known universe (Ω, in Figure 8-9). Trying to find the best arrangements of nodes and links in this huge space of possibilities can be quite exhaustive. There is some relief, however, in that the adjacency matrices created by combat networks are in a class called "sparse matrices." This means that for the simplest complete combat network only a fraction of the 1,844,670 billion possible sub-networks are actually formed.

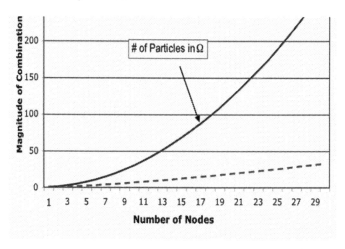

Figure 8-9. Network Dimensionality

DYNAMICS

An understanding of networked behavior comes from recognizing that dynamic behaviors are not found in static structure but result from the interactions of nodes over links. Specific arrangements of links and nodes that create value are "cycles," sub-networks in which the functions of nodes flow into each other over a path that revisits at least one node once. If there are no cycles in a network, then no useful networked function is completed. If there is to be advantage in using networked forces it must arise from these dynamic, often autocatalytic networked effects. Current NCW literature and contemporary combat models do not adequately describe these effects.

Not all collections of links and nodes, however, create cycles. Sub-networks with one or two nodes are not very robust or survivable as networks and would be rarer than fuller cycles. For example, a single sensor disconnected from a combat network is a sub-network of the larger set, but is not interesting from a combat network perspective. The same is true of a simple target-sensor pairing. These are known as 1- and 2-cycles, respectively. Three- and higher-dimension cycles contribute more to networked effects. In the Information Age Combat Model, there are of four general types of cycles.

Measuring Networked Effects

Various mathematical operations can be performed once a network has been converted to a matrix representation. A very rich and formal field of mathematics exists to perform these operations. One of the most useful operations is the calculation of eigenvalues. An eigenvalue, usually denoted by the Greek symbol λ, is a measure of the value of the network and is derived from the adjacency matrix.[40]

The adjacency matrices that describe the Information Age combat model are of a particular type, "sparse non-negative matrices," that have an important property that allows for measurement of networked effects. The Perron-Frebonius Theorem states that for matrices with this property, there exists at least one real non-negative eigenvalue larger than all others. In addition, since the entries in an adjacency matrix are 1's and 0's, the Perron-Frebonius eigenvalue (*PFE*) will have three distinct ranges of values which correspond to three distinct values of networked effects. The absence of a cycle, the presence of a simple cycle, and the magnitude of networked effects[5] (Jain & Sandeep, Graph Theory and the Evolution of Autocatalytic Networks, 2004).

The left side of Figure 8-10 shows a network without a cycle, indicated by the absence of a path from any node that returns to that node. The right side of the figure is the adjacency matrix that describes that noncyclical network. The *PFE* for the adjacency matrix is 0. By definition, an adjacency matrix with a *PFE* of 0 represents a network with no cycles.

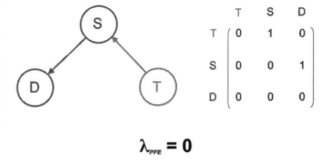

$\lambda_{PFE} = 0$

Figure 8-10. Network with No Cycles

---

[5] As with any multi-variant mathematical problem, there can be more than one eigenvalue that represents the value of a matrix.

Figure 8-11, by contrast, contains a simple cycle. The PFE of its adjacency matrix equals exactly 1. By definition, an adjacency matrix with a PFE of 1 represents a network with a simple cycle; a network with a simple cycle has no networked effects. Figure 8-12 shows network structure over and above the simple cycle in Figure 8-11. Such additional links and nodes add value to a network and are the mechanism by which networked effects accrue. The PFE of the matrix representing such an adjacency matrix measures the magnitude of networked effects and can be used to compare the topologies of various networks with respect to their potential for dynamic networked effects. These networks are called autocatalytic sets (ACSs) because the additional structure creates feed-forward and feedback linkages that autocatalytically create networked effects.

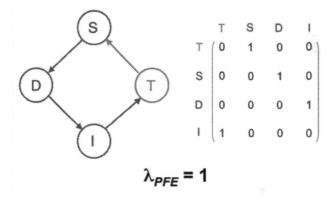

Figure 8-11. Network with a Single, Simple Cycle

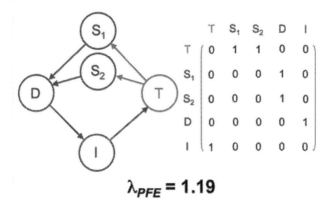

Figure 8-12. Network with an Autocatalytic Set (ACS)

Figure 8-13 shows how the PFE increases with additional linkage. Not all additional linkages, however, contribute to networked effects. Figure 8-14 for example, shows how the addition of a link and node to the basic structure in Figure 12 does not change the value of the PFE. The structure in Figure 8-12 is known as the "core" process of the network in Figure 8-13. A core is the set of links and nodes that contains all the mechanisms for networked effects in a network. Additional links and nodes that do not contribute to an increased PFE are called "peripheral" links and nodes. In larger networks, however, it is possible to have more than one core.

Since the largest possible PFE for an $N \times N$ adjacency matrix is N, then the networked effects of networks of different sizes can be compared using the ratio PFE/N, which we define here as the Coefficient of Networked Effects (CNE). CNE ranges in value from $1/N$ to 1.

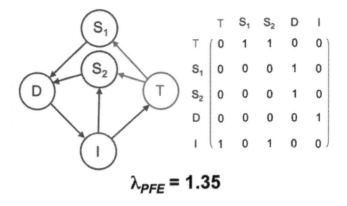

Figure 8-13. ACS with Additional Linkages

The examples presented here make it is clear that the characteristics of a network and its potential for improved performance increase as the network grows. The next section will discuss the importance of the long timescale dynamics of a network, or "network evolution."

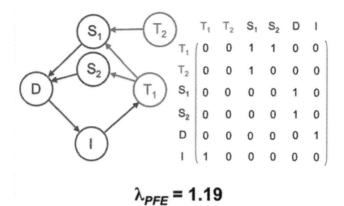

$$\lambda_{PFE} = 1.19$$

Figure 8-14. ACS with a Peripheral Link and Node

EVOLUTION

As networks mature they grow in a way that is very unlike progressive improvement in most other systems. This section defines and describes how exploitable properties develop because of growth and evolution in complex networks. A type of rapid connectivity ubiquitous in networked structures will be explored, mechanisms of adaptation and learning will be defined, and convergence toward a set of descriptive statistics will be discussed. The potential for using these statistics to quantify combat network performance will be addressed.

Punctuated Growth in Complex Networks

One of the most important phenomena in network evolution is punctuated growth. This pattern of sudden connectivity occurs as a network matures (under competition, for purpose or in response to resource constraint) from a loose collection of a small number of nodes into a larger, more complex structure.

A simple thought experiment demonstrates the essence of this rapid growth. Imagine that there are 400 buttons and many pieces of string on a table. Imagine also that a button and a piece of string are randomly selected from the table, tied together and placed back on the table. Now imagine this process is repeated indefinitely. Eventually, a button might be selected that has a string and a button already tied to it; perhaps also a string will be chosen that was previously connected to a button. Soon the table will be populated with so many clusters of buttons and strings that at some threshold level, adding a very few additional buttons or strings will connect almost all the small clusters into one large collection (called the giant component of the network).

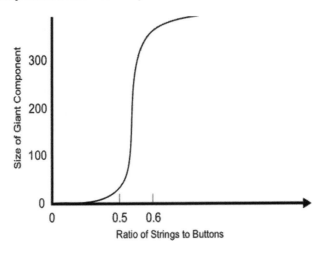

Figure 8-15. Buttons and Strings

Plotting the size of the giant component against the ratio of strings to buttons more formally describes punctuated growth. The curve in Figure 8-15 shows that as the ratio of strings to buttons approaches 0.5, the connectivity of the buttons suddenly and dramatically increases. The curve flattens quickly after this ratio hits 0.6, however, and each additional string adds only marginally fewer buttons to the network. Obviously, a connected network is not guaranteed by this method (the curve is asymptotic to the maximum number of buttons) but the method clearly displays the nature of the rapid transition from an unconnected group of nodes to a highly connected network (Kauffman, 1995). Although the buttons and strings, in this example, were connected in a purely random process, other complex networks experience the same type of punctuated growth. In addition, they experience the same "S-shaped curve is found in the growth profiles of all complex networks.

Learning and Adaptation in Complex Networks

Readers familiar with calculus-based engineering problems might look upon the curve in Figure 8-15 and find it quite familiar. With this traditional view, they would be mistakenly assume that the important system behaviors were occurring at the "knees" of both curves (where the curves "tip") and at the midpoint of the "S." There is latent structure in the part of the connectivity curve to the left of the "tipping point" at 0.5, however, that is far more important than the tipping points themselves. This latent structure is contained in the smaller clusters of nodes that eventually connect at the tipping point, but the tipping point will not occur unless this latent structure is present. In some networks, latent structure may account for up to 95% of connectivity, yet this substantial level of connectivity would not be measured in Figure 8-15 until a very large giant component is formed.

In complex networks this tail in the connectivity curve represents two distinct behaviors. The first behavior is a kind of learning, in the sense that the first small clusters of links and nodes inform the placement and connection of subsequent links and nodes (particularly in a combat network with sensors, but also in other networks that interact with the environment or a competitor). As additional links and nodes are added, the network evolves from one with no cycles to one with multiple simple cycles, and finally to one with ACSs and complex networked effects (Jain & K. Sandeep, Graph Theory and the Evolution of Autocatalytic Networks, 2004).

The second behavior is adaptation. When the environment or competition changes substantially, the arrangement of links and nodes and, therefore, the networked effects can become irrelevant to the competition or environment until such time as feedback or feed-forward results in reconfiguration of the network for its new relevant purpose. While subtlety distinct from the first behavior, reactive-learning, adaptation exploits the presence of latent links to help the network morph smoothly in response to environmental or competitive change. This adaptive reconfiguration can be achieved in complex network with a re-wiring of only 5-10% of the links.

A simple chain of links and nodes cannot be a complex network; a complex network, however, can invoke simple chains within it. Complex networks adapt, therefore by re-wiring simple chains with links selected out of latent structure. The latent structure is called "neutral" structure because it does not typically contribute to networked effects until it is incorporated into a re-wiring.

A measure of adaptability is the amount of latent structure—the amount of neutrality—in a complex network. This can be measured by subtracting the number of links in a simple chain of size N, $N-1$, from the number of links, l, in a network of size N. Dividing by N normalizes this calculation for network size and produces a statistic called the Neutrality Rating, $(l-N+1)/N$.

Core Shifts in Complex Networks

Learning or adaptation profoundly affects the dynamical structure of complex networks, particularly the dynamic relocation of the cores of networked effects. In a "core shift," the central mechanisms of networked effects move from one subset of links and nodes to another. An example of core shifts in the Information Age combat model follows. It portrays a combat network evolving from sensing a group of targets to attacking those targets.

Figure 8-16 shows a decision node controlling a group of sensors that detect a target. The core of the network is outlined by a box and the core portion of the adjacency matrix is highlighted by shading. The portion of the network outside the boxes and shading represents the presence of the two peripheral nodes (I_1 and I_2), as well as a target node (T).

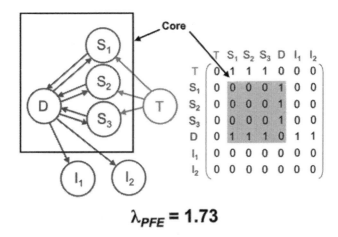

$\lambda_{PFE} = 1.73$

Figure 8-16. A Core of Sensors

Figure 8-17 shows the network adapting to sensors information by including two influencers in the coordination. Note that the core has expanded to include the influencers, and the *PFE* has changed as a result. Also note that this was accomplished by rewiring two control links from the sensors to the influencers and by invoking paths through two previously neutral links between the influencers and sensors. One sensor is now providing updated targeting information and the decision node has directed placement of the influencers.

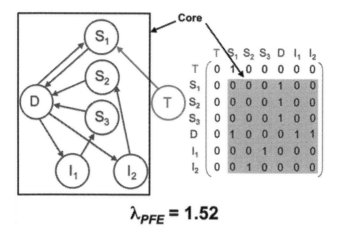

$\lambda_{PFE} = 1.52$

Figure 8-17. A Core of Sensors and Shooters

In Figure 8-18, the network has initiated an attack on T. The core has shifted and is now represented in the lower right corner and along the top and left side of the adjacency matrix. Sensors S_1 and S_2 have been re-allocated to search for additional targets and longer have a role in the attack. S_1 and S_2 are now peripheral to the network but, most importantly, T is in the core. Again, the PFE has changed with this shift in the core.

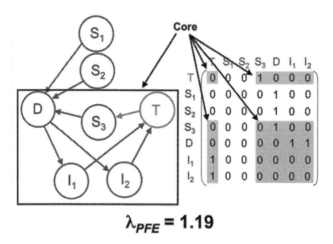

$\lambda_{PFE} = 1.19$

Figure 8-18. Attacking with a Core

The attack continues in Figure 8-19: the influencers continue to engage T and their progress is reported by sensor S_3. S_3 communicates data to the decision node D, which in turn applies additional control measures to I_1 and I_2. Note that while the underlying structure of the adjacency matrix has not changed (i.e., there has been no further core shift), the additional network interactions have resulted in a new *PFE*.

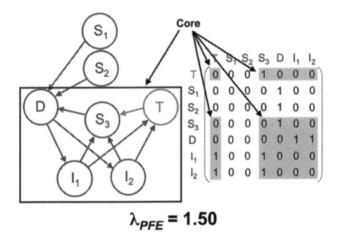

$\lambda_{PFE} = 1.50$

Figure 8-19. Attack by the Core Continues

Other Network Models

Neural Network Design

Training the neural network will be the most time consuming phase of the study and will require that a simulation model first be designed. The training process will operate in the following manner.

I. Design the neural network with fourteen input nodes, one bias node and one output node that represents the expected time to reach the destination. The appropriate number of hidden nodes will be determined by applying common rules of thumb (i.e. number of input nodes and number of output nodes divided by 2). The input and output nodes are defined as seen in the figure.

II. Build a simulation model to represent a hypothetical computer network. Two details will remain constant throughout the simulation process: the network topology and the source node of interest.

III. A large number of simulations will be run to develop a sufficient training set. Each simulation will consist of transmitting a message from a source node along each of the possible outgoing links using a data profile. Each profile for node one will consist of the following information:

  A. destination node
  B. packet size
  C. distance from the current node to the destination via the link in question
  D. throughput on the link
  E. number of packets traveling on that link
  F. type of failure present on that link

Physics-Based Communication Networks

Packet-level discrete-event network simulators use an event to model the movement of each packet in the network. This results in accurate models, but requires that many events are executed to simulate large, high bandwidth networks. Fluid-based network simulators abstract the model to consider only changes in rates of traffic flows. This can result in large performance advantages, though information about the individual packets is lost making this approach inappropriate for many simulation and emulation studies.

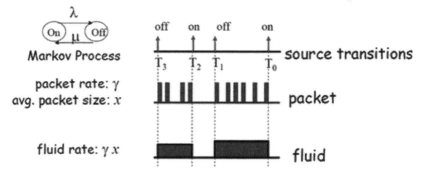

Figure 8-20. Two models of physics based communication networks

Packet-based model:
- network traffic flow: model packets in flow
- # sources, data rates increase, so too does simulation workload

Fluid-based model:
- network traffic flow: continuous fluid
  - rate changes at discrete points in time
  - rate constant between changes
- can modulate rate at different time scales
  - single modeling paradigm for many time scales
  - abstract out fine-grained details: simulation efficiency

Figure 8-21. Description of two models of physics based communication networks

A hybrid model, in which packet flows and fluid flows coexist and interact, may also be considered. This enables studies to be performed with background traffic modeled using fluid flows and foreground traffic modeled at the packet-level. Studies using the hybrid approach show up to 20 times speedup using this technique. Accuracy is within 4% for latency and 15% for jitter in many cases (Kiddle, Simmonds, Williamson, & Unger, 2003).

Virtual Cell Layout (VCL)

In VCL, the communications area is tessellated with virtual cells, which are fixed size hexagons that are placed starting from a reference geographic location. If an access point knows its geographic location, this location information can be mapped into a VCL cell index. This index can be used to determine the radio resources, which are a carrier set, and the code division multiple access (CDMA)[6] codes assigned to the VCL cell.

---

[6] Code division multiple access (CDMA) is a channel access method utilized by various radio communication technologies.

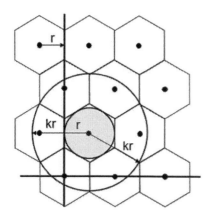

- The real cells are mobile and created by the mobile base stations, which are either:
  - radio access points (RAPs) or
  - cluster head man packed radios (MPRs).
- Computer aided exercise interacted tactical communications simulation (CITACS)
- A scenario with 153 units are simulated over an area of 115 km × 170 km
- Location manager deployed 77 RAPs and 18529 MPRs for this scenario based on the unit types and sizes.

Figure 8-22. Example of a virtual cell layout

The real cells are mobile and created by the mobile base stations, which are either radio access points (RAPs) or cluster head man packed radios (MPRs). The size of the real cells may be different from the size of the VCL cells. If the side length of a VCL cell is r, then the real cell radius becomes kr. We call k as the multiplication factor. When the multiplication factor is one, a real cell usually cannot cover the entire virtual cell where it is located because access points are not necessarily at the center of the virtual cells.

CITACS was used to evaluate the performance of these algorithms and schemes. CITACS interacts with Joint Theater Level Simulation (JTLS) and uses its outcomes in the detailed mobility, call and availability models for the tactical communications simulation.

This page internionally left blank

Chapter 9. Engagement Modeling – Entity Level

Engagement modeling has always been of primary importance in representing combat, both in virtual and constructive simulation. This modeling usually requires probabilities for various engagement aspects, including hit/miss and kill/no kills. However, many customers require more than just a determination of kill, e.g., amount of damage, mobility kill only, or some other degradation of combat capability. Although not discussed in detail here, many models still lack modeling of lethality after a engagement is performed.

For example, in Janus v. 7.1, individual soldier or individual system is the lowest entity modeled. Conventional direct fire from both ground and air systems is automatic and dependent on line-of-sight, probability of acquisition, identification and firing criteria, response time, reload rates, range, and posture of firer and of the target.

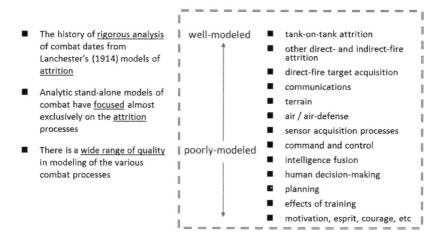

Figure 9-1. The trends of attrition modeling in combat models

The Role of Attrition in Combat Models

Engineering Models

GOAL: Determine how accurate and what damage a given shot does to a given target, i.e., tabulate bias and dispersion, PH|S and PK|H, for all shooter/target pairings.

High-Resolution Models

GOAL: Determine what damage a given firing weapon does to a given target.

Medium-Resolution Models

GOAL: Determine how many vehicles by type in a given unit are damaged by a given firing unit.

Low-Resolution Models

GOAL: Determine how a given target unit's effectiveness measure is degraded by participation in a battle.

The picture below represents the attrition process between military units (e.g., Blue tank platoon) in aggregated combat models. Inside the "Attrition" box is the attrition process for the disaggregated weapon systems (e.g., M1A2 Abrams), discussed in the next chapter. The smaller box that contains the "accuracy assessment" and the "damage assessment" is the "Physical Attrition Process", which we discuss in this chapter.

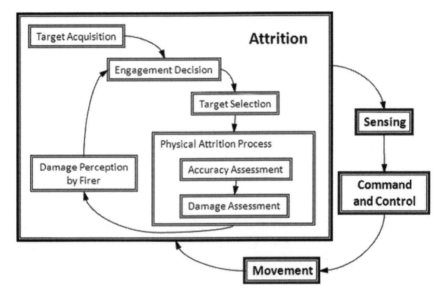

Figure 9-2. The role of attrition modeling in combat models

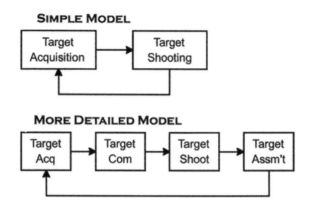

Figure 9-3. Serial Architecture Attrition Process

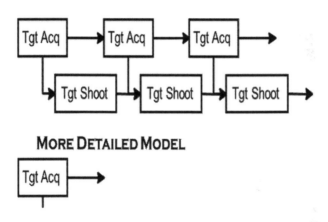

Figure 9-4. Local Parallel Architecture Attrition Process

Point System

Many shooter games use a simple point system to determine combat outcomes. Each target has a number of health points. Each weapon delivers a number of damage points. When a target's health point go below zero, it is dead.

There can be modifications to this. Armor, first aid, or other artifacts that make it stronger may supplement a target's health. These artifacts may be accounted for separate from health. This makes it possible to limit their effectiveness to certain types of weapons or attacks. For example, a flak vest may reduce the effect of a bullet to the chest, but have no effect on a direct hit by a fragmentation grenade launched from a weapon.

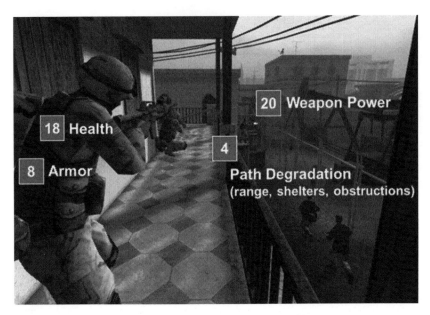

Figure 9-5. Screen shot from America's Army with hypothetical point system overlaid

$$New\ Health = (Health + Armor) - (Weapon\ Power - Path\ Degrade)$$

$$New\ Health = (18 + 8) - (20 - 4) = 10$$

$$New\ Armor = Armor - ABS[(Weapon\ Power - Path\ Degrade) * 0.25]$$

The damage points of a weapon may also be degraded based on environmental factors like range to target and obstacles along the path. This makes some weapons, like a shotgun, less effective at longer distances. It also allows for variation of effects when a grenade is thrown into a group or when the fragment pattern is partially blocked by walls.

Markov P_k Table

This method assigns a probability of kill (P_k) to specific weapon-target pairs. The data is generally stored in a Weapons-Target Matrix as shown in the picture. P_k values are assigned to each pair and can be adjusted as needed.

Though this method is very flexible, it contains two major weaknesses. First, the P_k values do not account for other variables such as range to target, closing speed, defensive positions, etc. Therefore, the values are often modified for each engagement before they are applied to the target.

P_k	Weapon				
	W1	W2	W3	W4	...
T1	0.5	0.7	0.8	0.92	
T2	0.4	0.45	0.76	0.99	
T3	0.31	0.34	0.56	0.85	
T4	0.27	0.55	0.67	0.81	
⋮					

Figure 9-6. Example of a Markov P_K table

Damage = 1, where Random Number ≤ P_k

= 0, where Random Number > P_k

Second, the values are all independent. It is possible to set any one value without regard for those around it. Therefore, it is difficult to insure consistency among the hundreds of numbers in the matrix. It is easy to tweak one number and unintentionally make a tank more vulnerable than a personnel carrier. This effect is compounded by the practice of applying modifiers to these numbers (for range, health, weather, etc.).

Random Numbers

When using Probabilities in engagement and attrition algorithms, random numbers suddenly become a core piece of the solution to the problem. Strictly, a Random Number falls between 0 and 1 and is evenly (or uniformly) distributed between these. The table in Figure 9-7 was generated using the RAND function in Microsoft Excel. For the most common cases, the techniques for generating good random numbers are well known. However, the mathematical approach used to create these numbers is less than perfectly uniformly distributed between 0 and 1. Therefore, all RAND functions are strictly referred to a Pseudo-random number generators.

If you need to select numbers for a distribution other than uniform, such a Normal, Poisson, Weibel, etc., you require random variates. The methods most often used to generate these variates (e.g., inverse transform) are described in many simulation textbooks, but will not be explored here. See (Strickland, Discrete Event Simulation Using ExtendSim 8, 2010).

0.920405	0.180168	0.139326	0.790158	0.255999	0.244471	0.616381
0.632651	0.017041	0.72307	0.023097	0.425282	0.868271	0.22006
0.391725	0.089067	0.647952	0.841486	0.211827	0.595938	0.163097
0.304131	0.593488	0.880553	0.913324	0.649363	0.933972	0.442742
0.18469	0.899123	0.115036	0.332234	0.652497	0.010936	0.560009
0.160788	0.427734	0.702527	0.041835	0.824957	0.237403	0.211855
0.959896	0.177809	0.184378	0.579104	0.95026	0.69267	0.057165
0.721817	0.502237	0.711678	0.181223	0.796627	0.663598	0.499042
0.47957	0.916742	0.14467	0.119294	0.234073	0.233899	0.301055
0.822892	0.603442	0.721817	0.291548	0.653141	0.709717	0.323901
0.930613	0.405974	0.123223	0.064081	0.314732	0.079808	0.159053
0.889581	0.694184	0.416254	0.217007	0.050884	0.806469	0.075488
0.604487	0.638116	0.812076	0.341242	0.973444	0.24498	0.185389
0.189308	0.886565	0.139973	0.001905	0.309989	0.963827	0.083722
0.307086	0.54353	0.151571	0.27651	0.845231	0.87254	0.429155

Figure 9-7. Table of Random Numbers

Pk's and Random Numbers

The probability of kill indicates the percentage of the time that the given event will result in the destruction of the object being attacked. In Figure 9-8, we show a P_k of 0.75 for a tank when attacked by some weapon. The random number selected was 0.63. This number falls between 0 and 0.75., therefore, the algorithm will decide that this vehicle is destroyed. If the random number had been greater than the P_k value, then the algorithm would have decided that the vehicle was not killed.

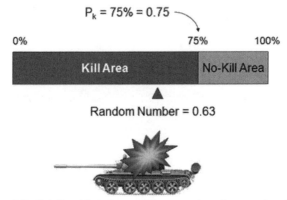

Figure 9-8. Relationship between kill area and random number draws

In this example, a single random number is used to make a kill/no-kill decision. In practice there are often many modifications to this to allow different types of kills based on the type of weapon, target, angle of engagement, and a number of other variables.

Precision Engagements

Automated crews could potentially achieve perfect weapon alignment on the target. However, these models must explicitly add errors to degrade their performance to some determined human level. Each target provides some presented area to the shooter. The shooter places an aim point on this vehicle and fires the weapon. However, the engagement is modified by a fixed bias for the type of weapon used (AMSAA or JMEMS Tables), an occasional bias for the variations caused by the positioning of two vehicles, and a random bias for the round-to-round differences found in any form of munition.

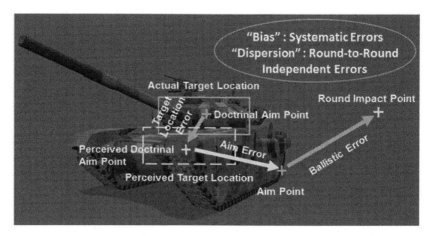

Figure 9-9. What happens in a simulation of a precision engagement? (Strickland, Mathematical Modeling of Warfare and Combat Phenomenon, 2011)

These model effects prevent automated forces from performing perfectly. The picture shows the application of three different errors. This could require three random numbers ad each factor may differ from one engagement to the next, or can be simplified into a single random draw if the three values do not change. Precision engagements were implemented in Mod SAF.

Linear Target P_{hit}

Recall that the normal probability distribution, with mean μ and standard deviation σ, is expressed by

$$p(x) = \frac{1}{\sqrt{2\pi}\sigma} \exp\left[-\frac{(x-\bar{x})^2}{\sigma^2}\right]$$

The area under the normal curve between x_1 and $x = \infty$ is given by

$$P(x_1 < x < \infty) = \int_{x_1}^{\infty} p(x)dx$$

It is generally useful to compute these probabilities in terms of u standard deviations from the mean value $\bar{x}$ as defined by $x = \bar{x} + u\sigma$ or $u = (x - \bar{x})/\sigma$. A change in variable x into u with $dx = \sigma du$ and substitution of (**1.1**) into (**1.2**) leads to

$$P(u_1 < u < \infty) = \frac{1}{\sqrt{2\pi}} \int_{u_1}^{\infty} e^{-\frac{1}{2}u^2} du$$

which is independent of the standard deviation σ.

In most statistical work and in the analysis of navigational errors, the convenient measure of error is the standard deviation, but in weapon effect analysis we generally use the 50% error, or the so-called **error probable**. This implies that the tail areas on either side of the mean value $\bar{x}$ must be equal to 0.25 as shown Figure 9-10. This is the linear error probable or probability of 50% error. Using a table of normal probabilities, it follows that for this case $u = \pm 0.6745$. And because the total area under the curve is equal to unity, the central area is therefore equal to 0.50 and defines the **linear errorprobable** (LEP). Hence, using the definition for u, the LEP in range is given by

$$LEP_{range} = 0.6745\sigma$$

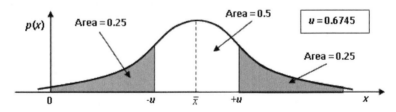

Figure 9-10. Normal distribution showing 50% probability in the tails, or Linear error probable or probability of 50% error.

Example. For a particular tank gun the range error probable is 25 meters (m). We want to find the probability of hitting a target 20 m long when aiming at the center of the target. We will assume that there is no systematic error and the normal distribution is centered around the aiming point, as in Figure 1-3. In this case, $\bar{x} = 0$, $\sigma = 25/0.6745 = 37.0644\ m$, $x = \pm 10\ m$, and $u = \pm 10/37.0644 = \pm 0.2698$. Using a table of normal probabilities, $P(u > 0.2698) = 0.3937$. Hence, the probability of hitting the target is

$$P(-0.2698 < u < 0.2698) = 1 - 2(0.3937) = 0.2127$$

This is the single-shot probability of hitting the target, denoted PSSH.

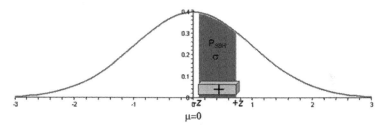

Figure 9-11. Normal distribution for example of single shot accuracy

This value represents the single-shot probability of hitting the target and is usually denoted P_{SSH}. The probability of a miss is therefore $(1 - P_{SSH})$. If the target is fired upon n times, then the probability of a miss is $(1 - P_{SSH})^n$ and the probability of a hit is

$$\boldsymbol{P_n = 1 - (1 - P_{SSH})^n}$$

Figure 9-10 shows the single-shot probability for a linear, one-dimensional target. The algorithm assumes that the firer is aiming at the center of the target and that error occurs because the weapon is not perfect. This round-to-round error in dispersion is assumed to be a normally distributed random variable.

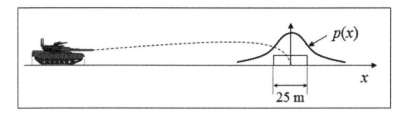

Figure 9-12. Probability distribution for a direct fire engagement in one dimension

Normal parameters for 1D target:
- "Front view" (i.e., direct-fire weapon)
  - Deflection error
- "Top view" (i.e., indirect-fire weapon)
  - Range error
- DEFINE:
  - Bias = μ
  - Dispersion = σ

Error probable is an important concept in probability of hit algorithms, and is often mistakenly considered a probability. The correct interpretation of error probable is the distance in deflection within which half of the rounds will land.

Linear error probable (LEP) is the linear distance from the aim point within which half of the rounds will land, based on the error probable.

Single-Shot Accuracy

Assume no systematic error and that the normal distribution is centered around the target aiming point. For a particular tanks gun, the range error probable is 25 m. We want to find the probability of hitting a target 20 m long when aiming at the center of the target.

$$\mu = 0, \sigma = 25/0.6745 = 37.0644 \text{ m}, x = \pm 10 \text{ m},$$

then $z = \pm 10/37.0644 = \pm 0.2698$

Using a table of probabilities, $P(z > 0.2698) = 0.3937$, hence the probability of hitting the target is

$$P(-0.2698 < z < 0.2698) = 0.6063 - 0.3937 = 0.2127$$

Rectangular Target P_{hit}

The ***range error probable*** (REP) and cross-range error probable (CREP) are equivalent to the LEP (in one dimension). The ***circular error probable*** (CEP) is a function of one parameter, radius r, obtained through the transformation of the bivariate normal distribution, with $r = \sqrt{x^2 + y^2}$ (Strickland, Mathematical Modeling of Warfare and Combat Phenomenon, 2011).

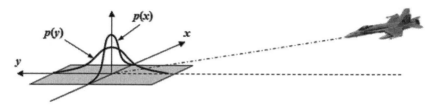

Figure 9-13. Probability distribution for a direct fire engagement in two dimensions (CREP and REP)

Normal parameters for 2D target:
- "Side view" (i.e., direct-fire weapon)
  - Elevation error
  - Deflection error
- "Top view" (i.e., indirect-fire weapon)
  - Range error
  - Deflection error
- DEFINE:
  - Bias = μ_x, μ_y
  - Dispersion = σ_x, σ_y

Circular Target P_{hit}

The probability of destruction of a point target for a cookie-cutter damage function with lethal radius R is the same as the probability of hit within a circle of radius R, i.e., Pd = P.

$$P(R_0) = 1 - exp\left(-\frac{R^2}{2\sigma^2}\right) = \frac{1}{2}$$

A companion concept to the range and cross-range probable error or the 50% error is the circular error probable. If r denotes the radius of a circle for which then the meaning of such a radius is that 50% of all the impact points for the probability of destruction $p(x, y) = p(r)$ will fall within this radius, $r \leq R_0$, and the remaining 50% will fall outside, . This radius is the circular error probable (CEP) and plays a very important role is assessing the accuracy of any weapon system (Strickland, Mathematical Modeling of Warfare and Combat Phenomenon, 2011).

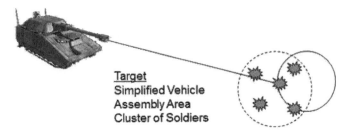

Figure 9-14. Direct fire with circular target area (CEP)

Kill Categories

Once mutually exclusive values are obtained, it is a simple matter to select the kill effects using a single random number draw that is applied to the kill categories. Simplifications of this sort are often performed on empirical data that is used in simulations. The form in which the data is collected is not necessarily the form best used in the simulation.

Figure 9-14 shows examples of kill categories. Notice that we have attempted to stack the P_k's from least lethal to most lethal. If this can be done, then it is easier to apply a P_k modifier directly to the P_k and continue using this equation. Examples of modifiers are night, rain, suppressing fire, fatigue, etc. Multiple kill categories are implemented in Janus v. 7.1.

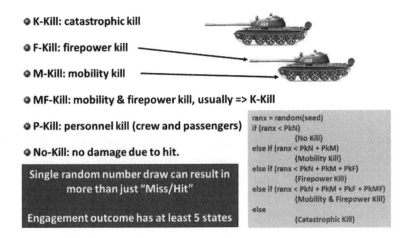

Figure 9-15. Examples of kill categories and algorithm

Direct-Fire Accuracy Example

A direct fire model can be implemented as a Monte Carlo simulation. The frontal profile of an infantry fighting vehicle (IFV) is depicted above, with regions representing area that ,if hit, result in different kill categories. We can model the direct fire frontal shot impact point on the IFV as a bivariate normal random variable distributed as $(X, Y) \sim BVN(0, 0, 0.5, 0.5, 0)$. The last parameter in this distribution is the covariance. We then generate a set of random numbers to use in the simulation (this can be accomplished using the RAND function in MS Excel).

The details of the Monte Carlo simulation are shown here, and uses Normal probability lookup tables. However, this is easily implemented in MS Excel. Given the set of random numbers used, the target is hit in area 4, a hit that will produce no permanent effect.

An infantry-fighting vehicle (IFV) has the following frontal profile (Strickland, Mathematical Modeling of Warfare and Combat Phenomenon, 2011):

- A hit in area 1 will produce a firepower kill.
- A hit in area 2 will produce a catastrophic kill.
- A hit in area 3 will produce a mobility kill.
- A hit in other areas will produce no permanent effect.

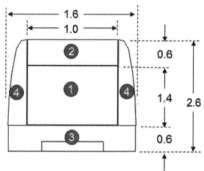

Assess the IFV's vulnerability when engaged with a frontal shot whose impact point is modeled as a random variable pair $(X, Y) \sim BVN(0,0,.5,.5,0)$.

Using the below list of pseudo random numbers as needed, simulate the first round to determine which type of kill, if any, occurs (.8554, .2287, .6659, .8243, .6840, .0430, .8598, .2381, .5035, .2723).

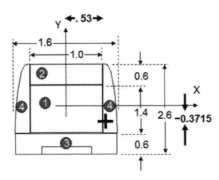

1. Do a Monte Carlo simulation of impact point with origin centered on the target, then compare impact point with target profile to calculate where it hit.
2. Determine X coordinate of impact point:
   - Enter the Normal Table with 0.8554
   - Find $Z^{-1} = 1.06$
   - Note that $Z^{-1} = ((x - \mu_x)/\mu_x$
   - Solve for x in $1.06 = (x - 0)/0.5$
   - $x = 0.53$

3. Determine the Y coordinate of the impact point (using RN .2287):

- Normal Table goes from 0.5000 to 0.9999, but Normal Dist. is symmetric, so compute $1.0 - 0.2287 = 0.7713$, and change sign of resulting Y coordinate.
- Interpolating between 0.75 and 0.74, gives $Z^{-1} = 0.743$.
- Solve for y in $-0.743 = (y - 0)/0.5$ gives $y = -0.3715$

4. Round hits area 4, so no kill is assessed.

Chapter 10. Engagement Modeling – Aggregate Level

Combat process descriptions for these smaller echelons are radically different from the high-resolution event sequences. Since the model does not keep track of individual attributes, it cannot know details of what a particular individual is doing at any time. Instead, aggregated models represent the average results of many combatants interacting over a period of time by using the rates at which various process outcomes occur. As a simple example, an attrition equation might compute enemy casualties during a time interval as

$$Y\_CASUALTIES = X\_FIRERS * ATTR\_RATE * \Delta T$$

where X_FIRERS is the average number of friendly shooters in the battle, $ATTR\_RATE$ is the average rate at which a single surviving friendly shooter kills enemy systems, and ΔT is the length of the engagement.

Aggregated attrition process models can be categorized into two basic types that correspond to the two basic entity aggregation patterns. homogeneous and heterogeneous. In a homogeneous attrition model, combat attrition is assessed against a scalar measure of the unit's combat power. A heterogeneous attrition model assesses combat attrition caused by weapon system classes against enemy weapon system classes within the combat units.

Aggregate level engagement modeling has many issues that we do not have perfect solutions for, e.g., going from aggregate to disaggregate and vice versa, and involves consistency, estimation of process coefficients, and calibration.

Lanchester-type differential equations, aggregated combat groups (or piston models), and force ratio attrition models are some examples of aggregate level attrition models.

Lanchester Equations

History: The British engineer F. W. Lanchester (1914) developed this theory based on World War I aircraft engagements to explain why concentration of forces was useful in modern warfare. The original formulation was in terms of two models, called the square law and the linear law (or Lanchester equations). Many extensions and modifications have been proposed to add robustness and detail. (JANUS, CASTFOREM, and VIC uses modifications.)

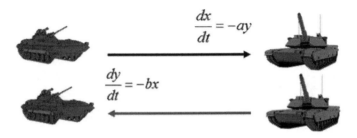

Figure 10-1. Tank (unit) versus tank (unit) direct fire attrition model

This model underlies many low-resolution and medium-resolution combat models. Similar forms also apply to models of biological populations in ecology. Assume that the combat casualty rate for the X force is proportional to the strength of the Y force. We will ignore all other factors in our initial model so that the rate of change in is

$$\frac{dx}{dt} = -ay, \quad a > 0$$

The positive constant a in this equation is called the antitank weapon kill rate or attrition-rate coefficient and reflects the degree to which a single antitank weapon can destroy tanks. The simplest assumption is that the loss rate is proportional to the number of firers.

Under similar assumptions, the rate of change in the Y force is given by

$$\frac{dy}{dt} = -bx, \quad b > 0$$

Here the constant b indicates the degree to which a single tank can destroy antitank weapons. The autonomous system given by these two equations, together with the initial strength levels, is called a Lanchester-type combat model (Strickland, Mathematical Modeling of Warfare and Combat Phenomenon, 2011).

Differential Equations

A differential equation is an equation involving one or more derivatives of the dependent variable, e.g., dx/dt.

$$\frac{dx}{dt} = -ax.$$

Other notation include

$$x'(t) = -ax$$

$$\dot{x}(t) = -ax.$$

Solutions to the Lanchester Equations

A solution to a differential equation is a function that satisfies the differential equation.

For example, the function $x(t) = e^{-\alpha t}$, satisfies differential equation (1):

$$\frac{d}{dt} e^{-\alpha t} = -\alpha e^{-\alpha t},$$

where

$$\frac{d}{dt} e^{-\alpha t} = \frac{d}{du} e^u \cdot \frac{d}{du} u = -\alpha e^{-\alpha t}.$$

The basic solution technique for this form of differential equation is called separation of variables.

Steps:

Separate variables to different sides of the equality

Integrate each side with respect to the variable of that side, within a constant

Simplify each expression

$$\frac{dx}{dt} = -\alpha x$$

$$\frac{1}{x} dx = -\alpha dt$$

$$\int \frac{1}{x} dx = \alpha \int dt$$

$$\ln x = \alpha t + C$$
$$x(t) = e^{-\alpha t + C} = e^C e^{-\alpha t} = K e^{-\alpha t}$$

Initial Value Problem

Then, if

$$x(0) = x_0, \quad \text{then}$$
$$x(0) = Ke^{-\alpha \cdot 0} = Ke^0 = K$$
$$\Rightarrow K = x_0$$
$$\text{So}, x(0) = x_0 e^{-\alpha t}$$

Systems of DiffEq's

A system of differential equations with two independent variables:

$$\frac{dx}{dt} = -\alpha y \text{ and } \frac{dy}{dt} = -\beta x$$

with $x(0) = x_0$ and $y(0) = y_0$. This system is called a first-order linear system (Strickland, Fundamentals of Combat Modeling, 2011).

A set of solutions that satisfies these this system is:

$$x(t) = x_0 \cosh \sqrt{\alpha \beta} t - y_0 \sqrt{\alpha/\beta} \sinh \sqrt{\alpha \beta} t \tag{3}$$

$$y(t) = y_0 \cosh \sqrt{\alpha \beta} t - x_0 \sqrt{\beta/\alpha} \sinh \sqrt{\alpha \beta} t \tag{4}$$

Linear Systems

We can write a system of linear differential equations in matrix form:

$$\frac{d\vec{Y}}{dt} = \vec{A}\vec{Y}$$
$$\begin{bmatrix} dx/dt \\ dy/dt \end{bmatrix} = \begin{bmatrix} 0 & -\alpha \\ -\beta & 0 \end{bmatrix} \begin{bmatrix} x \\ y \end{bmatrix},$$

where $\vec{A}$ is an $n \times n$ coefficient matrix and $\vec{Y}$ is a vector of independent variables.

The right-hand side of this equation is a vector field, $V(x, y)$; that is, a function that assigns a vector to each point in the xy-plane.

Eigen-Solutions

A first-order linear system has a straight-line solution.

That is, there is a vector $\vec{V}$ such that $\vec{A}\vec{V}$ points in the same or in the opposite direction as the vector from $(0,0)$ to (x,y), for which there is some number λ such that

$$\vec{A}\vec{V} = \lambda \vec{V}$$

Definition. Given a matrix $\vec{A}$, and number λ is called an eigenvalue of $\vec{A}$ if there is a nonzero vector $\vec{V} = \langle x, y \rangle$ for which

$$\vec{A}\vec{B} = \begin{bmatrix} a_{11} & \cdots & a_{1n} \\ \vdots & \ddots & \vdots \\ a_{m1} & \cdots & a_{mn} \end{bmatrix} \begin{bmatrix} v_1 \\ v_2 \\ v_3 \end{bmatrix} = \lambda \begin{bmatrix} v_1 \\ v_2 \\ v_3 \end{bmatrix} = \lambda \vec{V}$$

The vector $\vec{V}$ is called an eigenvector corresponding to the eigenvalue λ.

To find straight-line solutions of linear systems, we must find the eigenvalues and eigenvectors of the corresponding coefficient matrix, $\vec{A}$.

That is, we need to find the vectors $\vec{V}$, such that $\vec{A}\vec{V} = \lambda \vec{V}$. If

$$\vec{A} = \begin{bmatrix} 0 & -\alpha \\ -\beta & 0 \end{bmatrix},$$

then

$$\begin{bmatrix} 0 & -\alpha \\ -\beta & 0 \end{bmatrix} \begin{bmatrix} v_1 \\ v_2 \end{bmatrix} = \lambda \begin{bmatrix} v_1 \\ v_2 \end{bmatrix}.$$

We want to find nonzero solutions to

$$\vec{A}\vec{V} = \lambda \vec{V}$$
$$\vec{A}\vec{V} - \lambda \vec{V} = \vec{0}$$
$$(\vec{A} - \lambda \vec{I})\vec{V} = \vec{0}$$

where $\vec{I}$ is the $n \times n$ identity matrix.

The determinant condition for a nontrivial solution of the equation $\vec{A}\vec{V} = \lambda \vec{V}$ is

$$\det(\vec{A} - \lambda \vec{I})\vec{V} = \vec{0}$$

$\det(\vec{A} - \lambda \vec{I})\vec{V} = \vec{0}$ is called the **characteristic equation** or polynomial. If

To find the eigenvalues of the matrix $\vec{A}$, we must find the values of λ for which $\det(\vec{A} - \lambda \vec{I})\vec{V} = \vec{0}$.

$$\det(\vec{A} - \lambda \vec{I})\vec{V} = \det \begin{bmatrix} 0 - \lambda & -\alpha \\ -\beta & 0 - \lambda \end{bmatrix}$$

$$= (0 - \lambda)(0 - \lambda) - (-\alpha)(-\beta) = \lambda^2 - \alpha\beta = 0$$

$$\Rightarrow \lambda^2 = \alpha\beta \Rightarrow \lambda = \pm\sqrt{\alpha\beta}$$

Thus

$$\lambda_1 = \sqrt{\alpha\beta} \text{ and } \lambda_2 = -\sqrt{\alpha\beta}$$

To find a vector $\vec{V}_1$ for λ_1, we must solve

$$\vec{A} \begin{bmatrix} u_1 \\ v_1 \end{bmatrix} = \begin{bmatrix} 0 & -\alpha \\ -\beta & 0 \end{bmatrix} \begin{bmatrix} u_1 \\ v_1 \end{bmatrix} - \sqrt{\alpha\beta} \begin{bmatrix} u_1 \\ v_1 \end{bmatrix}$$

$$\left(\begin{bmatrix} 0 & -\alpha \\ -\beta & 0 \end{bmatrix} - \sqrt{\alpha\beta} \begin{bmatrix} 1 & 0 \\ 0 & 1 \end{bmatrix} \right) \begin{bmatrix} u_1 \\ v_1 \end{bmatrix} = \begin{bmatrix} 0 \\ 0 \end{bmatrix}$$

$$\begin{bmatrix} -\sqrt{\alpha\beta} & -\alpha \\ -\beta & -\sqrt{\alpha\beta} \end{bmatrix} \begin{bmatrix} u_1 \\ v_1 \end{bmatrix} = \begin{bmatrix} 0 \\ 0 \end{bmatrix}$$

Since $(\vec{A} - \lambda \vec{I})$ is a singular matrix, we let $u_1 = 1$, then

$$v_1 = -\frac{\sqrt{\alpha\beta}}{\alpha} \text{ and } \vec{V}_1 = \begin{bmatrix} u_1 \\ v_1 \end{bmatrix} = \begin{bmatrix} 1 \\ -\sqrt{\alpha\beta}/\alpha \end{bmatrix}$$

To find a vector $\vec{V}_2$ for λ_2, we must solve

$$\vec{A} \begin{bmatrix} u_2 \\ v_2 \end{bmatrix} = \begin{bmatrix} 0 & -\alpha \\ -\beta & 0 \end{bmatrix} \begin{bmatrix} u_2 \\ v_2 \end{bmatrix} + \sqrt{\alpha\beta} \begin{bmatrix} u_2 \\ v_2 \end{bmatrix}$$

$$\left(\begin{bmatrix} 0 & -\alpha \\ -\beta & 0 \end{bmatrix} + \sqrt{\alpha\beta} \begin{bmatrix} 1 & 0 \\ 0 & 1 \end{bmatrix} \right) \begin{bmatrix} u_2 \\ v_2 \end{bmatrix} = \begin{bmatrix} 0 \\ 0 \end{bmatrix}$$

$$\begin{bmatrix} \sqrt{\alpha\beta} & -\alpha \\ -\beta & \sqrt{\alpha\beta} \end{bmatrix} \begin{bmatrix} u_2 \\ v_2 \end{bmatrix} = \begin{bmatrix} 0 \\ 0 \end{bmatrix}$$

Since $(\vec{A} - \lambda \vec{I})$ is a singular matrix, we let $u_2 = 1$, then

$$v_2 = \frac{\sqrt{\alpha\beta}}{\alpha} \text{ and } \vec{V}_2 = \begin{bmatrix} u_2 \\ v_2 \end{bmatrix} = \begin{bmatrix} 1 \\ \sqrt{\alpha\beta}/\alpha \end{bmatrix}$$

Theorem. Suppose the matrix A has a real eigenvalue λ with associated eigenvector $\vec{V}$. Then the linear system $dY/dt = AY$ has the straight-line solution

$$\vec{y}(t) = e^{\lambda t} \vec{V}$$

Moreover, if λ_1 and λ_2 are distinct, real eigenvalues with eigenvectors V_1 and V_2 respectively, then the solutions $Y(t) = e^{\lambda_1 t} V_1$ and $Y(t) = e^{\lambda_2 t} V_2$ are linearly independent and

$$\vec{y}(t) = k_1 e^{\lambda_1 t} \vec{V}_1 + k_2 e^{\lambda_2 t} \vec{V}_2$$

is the general solution of the system.

The solution to our example is:

$$\vec{Y}(t) = k_1 e^{\sqrt{\alpha\beta} t} \begin{bmatrix} 1 \\ -\sqrt{\alpha\beta}/\alpha \end{bmatrix} + k_2 e^{-\sqrt{\alpha\beta} t} \begin{bmatrix} 1 \\ \sqrt{\alpha\beta}/\alpha \end{bmatrix}$$

Therefore

$$x(t) = k_1 e^{\sqrt{\alpha\beta} t} + k_2 e^{-\sqrt{\alpha\beta} t}$$

$$y(t) = -k_1 e^{\sqrt{\alpha\beta} t} (\sqrt{\alpha\beta}/\alpha) + k_2 e^{-\sqrt{\alpha\beta} t} (\sqrt{\alpha\beta}/\alpha)$$

Assuming the initial values $x(0) = x_0$ and $y(0) = y_0$, we can solve for k_1 and k_2 by

$$\vec{Y}(0) = k_1 e^0 \begin{bmatrix} 1 \\ -\dfrac{\sqrt{\alpha\beta}}{\alpha} \end{bmatrix} + k_2 e^{-0} \begin{bmatrix} 1 \\ \dfrac{\sqrt{\alpha\beta}}{\alpha} \end{bmatrix}$$

$$\begin{bmatrix} x_0 \\ y_0 \end{bmatrix} = \begin{bmatrix} 1 & 1 \\ -\dfrac{\sqrt{\alpha\beta}}{\alpha} & \dfrac{\sqrt{\alpha\beta}}{\alpha} \end{bmatrix} \begin{bmatrix} k_1 \\ k_2 \end{bmatrix}$$

$$\begin{bmatrix} k_1 \\ k_2 \end{bmatrix} = \begin{bmatrix} 1 & 1 \\ -\dfrac{\sqrt{\alpha\beta}}{\alpha} & \dfrac{\sqrt{\alpha\beta}}{\alpha} \end{bmatrix}^{-1} \begin{bmatrix} x_0 \\ y_0 \end{bmatrix}$$

$$\begin{bmatrix} k_1 \\ k_2 \end{bmatrix} = \begin{bmatrix} \dfrac{x_0}{2} - \dfrac{y_0}{2}\sqrt{\dfrac{\alpha}{\beta}} \\ \dfrac{x_0}{2} + \dfrac{y_0}{2}\sqrt{\dfrac{\alpha}{\beta}} \end{bmatrix}$$

Making a long story short

$$x(t) = \dfrac{x_0 e^{\sqrt{\alpha\beta} t} - y_0 \sqrt{\dfrac{\alpha}{\beta}} e^{\sqrt{\alpha\beta} t}}{2} + \dfrac{x_0 e^{-\sqrt{\alpha\beta} t} - y_0 \sqrt{\dfrac{\alpha}{\beta}} e^{-\sqrt{\alpha\beta} t}}{2}$$

$$= \left(\frac{x_0 e^{\sqrt{\alpha\beta}t} + x_0 e^{-\sqrt{\alpha\beta}t}}{2}\right) - \left(\frac{y_0\sqrt{\frac{\alpha}{\beta}}e^{\sqrt{\alpha\beta}t} - y_0\sqrt{\frac{\alpha}{\beta}}e^{-\sqrt{\alpha\beta}t}}{2}\right)$$

$$= x_0\left(\frac{e^{\sqrt{\alpha\beta}t} + e^{-\sqrt{\alpha\beta}t}}{2}\right) - y_0\sqrt{\frac{\alpha}{\beta}}\left(\frac{e^{\sqrt{\alpha\beta}t} - e^{-\sqrt{\alpha\beta}t}}{2}\right)$$

$$= x_0 \cosh\sqrt{\alpha\beta}t - y_0\sqrt{\frac{\alpha}{\beta}}\sinh\sqrt{\alpha\beta}t$$

This satisfies equation (3).

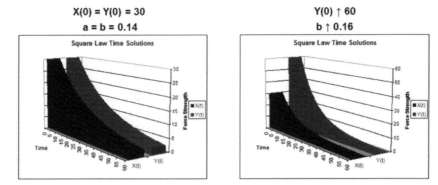

Figure 10-2. Differences in attrition between Lanchester models with different initial conditions. The chart on the right indicates that the initial size of forces may be a predominating factor. (See (Strickland, Mathematical Modeling of Warfare and Combat Phenomenon, 2011) for further details.)

Aggregated Combat Groups

Joshua Epstein at the Brookings Institute sought a very abstract level of representation for force-on-force engagements. He wanted to minimize the number of weapon-target calculations in order to arrive at solutions for very large theater level models. He created combat groups that came to be referred to as "pistons." These pistons travel in invisible tracks that cover the entire battlefield. Combat is calculated between the accumulated forces in each side of the piston. Units are more effective and more vulnerable if they move toward the front of the piston.

As attrition occurs, one side becomes weaker and the piston slides in that direction, representing success at pushing the enemy back. However, this results in the piston accumulating forces that were immediately behind it, strengthening the losing side, stopping the movement of the piston, and perhaps forcing it in the other direction.

One implication of this type of model is that, since the piston is the sole method for combat modeling, blue and red units must be prevented from encountering each other directly. This is usually done by applying "friction" factors which keep the pistons contiguous with each other, eliminating holes through which enemy units could pass.

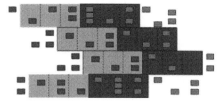

- Contiguous pistons
- Aggregated force attrition
- Distance from middle affects power and attrition
- Units accumulate as piston moves
- Explicit withdrawal required

Figure 10-3. A depiction of a typical piston model

Thunder is a simulation that uses pistons. In the THUNDER model, ground war elements are deterministic, based on a simple linear piston model. Attrition in the ground war is calculated using the Attrition Calibration (ATCAL) model developed by the U.S. Army Concepts Analysis Agency.

Force Ratio Attrition Models

The basic idea of homogenous force ratio attrition models is to aggregate all the individual combatants in a unit into a scalar measure of the unit's combat power (Parry & Hatman, 1992). The Firepower score approach is used in, aggregated, large-scale, combat simulations as the primary descriptor of what a combat unit is worth (Parry & Hatman, 1992). The ratio of attacker to defender combat power is used to determine the casualties for both sides.

CONCEPT:

> ➤ Summarize effectiveness in combat with a single scalar measure of combat power for each unit.
> ➤ When combat occurs, use the ratio of attacker's to defender's measures to determine the outcome

➢ Measures can be very subjective, or based on summary characteristics of unit equipment, or based on a weighted combination of weapon firepower, mobility, vulnerability, etc.

Force Ratio Approach - Firepower Scores

In the Firepower score approach, the combat power of a unit is computed by summing the combat power value for each weapon system in the unit. In Parry's notes (PARRY, 1992) the combat power computation is given in a simple equation as follows:

Suppose that there are n different types of weapon system in a combat unit and that:

X_i: the number of weapons of type i in the unit [i = 1,2,3…n]

S_i: the firepower score value representing the combat power for each type i weapon. Then, the firepower index of the aggregated unit is

$$\mathbf{FPI} = \sum_{i=1}^{n}(x_i s_i) = \text{firepower index of unit}$$

Finally, the force ratio is determined as:

$$\mathbf{FR} = \frac{FPI_{Attacker}}{FPI_{Defender}} = \frac{FPI(A)}{FPI(D)}$$

The force ratio gives a measure of relative combat power in the battle. The force ratio in many aggregated combat models, such as TACWAR, is used to compute casualties for both sides in a battle, and to determine the FEBA (forward edge of the battle area) or FLOT (forward line of troops) movement rates.

Force Ratio Approach - Limitations of Firepower Scores

The quality of the available historical data for validation of attrition models is very poor. The most accessible battle data contain only starting sizes and casualties and sometimes only for one side (Bracken, Kress, & Rosenthal, 1995). Recently, more data has become available. Improved database management and computing power have helped in gathering such data.

Detailed databases on the Battle of Kursk, the largest tank battle in history, and the Ardennes Campaign of World War II, have recently been developed. Both data sets are two-sided, time-phased (daily) and detailed. Hartley and Helmbod pointed out that unless we are able to procure time-phased detailed data, we will not be able to validate any attrition model (Hartley & Helmbold., 1995). Most of the past empirical validation studies have focused on the Lanchester Equations.

Unlike the Lanchester equations, there is no study in the literature that used firepower score attrition models on real data, in which force sizes are available day by day for both sides. The first attempt to validate the firepower score attrition models is conducted in Gozel's study (GOZEL, 2000)

Table X.X. Limitations of Firepower Scores

- Weapon system interactions ("synergisms") are not represented:
  - FPI is additive across weapon types (e.g., FPI=100 derived from 15 tanks, 30 infantry fighting vehicles and 6 mortars equals FPI=100 derived from 30 tanks).
- FPI is linear in the number of weapons - cannot directly represent:
  - Minimum unit size required for effectiveness.
  - Diminishing returns with large numbers of systems.
- Weapon system types become submerged in the FPI formula:
  - Illogical battles can occur (e.g., Arty Bn FPI=80 defeats Tank Co FPI=30).
- Validity is suspect because:
  - Specification of firepower scores is arbitrary
  - Force-Ratio predictive power is not validated by empirical historical research.
- Then why are we learning about Firepower Scores?
  - Because they match the rule-of-thumb approach often used by military planners.
  - Because you need to know enough about them to note when they are misused.

Force Ratio Approach - Determining Firepower Scores

Alternate Approaches:

There are several approaches to determining firepower scores for a force ration model. One way is to use historical data about combat performance (this was done in the 1950s). Another is to use the technical measures of weapon firepower (used in *ATLAS*). A third way is to use situationally-dependent technical firepower measures. Finally, one can use weighted combinations of firepower, mobility, vulnerability, reliability, etc., where weights were assigned by "Delphi" analysis (consensus of "experts"). This approach uses the term WEI/WUV (pronounced "wee-wuv")

WEI (Weapon Effectiveness Index): Weapon scores are given without regard to opposing target types

WUV (Weighted Unit Value):

- Units are given a base value by summing the WEIs of all their weapons
- Unit values are weighted by subjective ratings (e.g., for training, morale, C2 systems)
- Simple, but very subjective

Use a measure of what the weapon can kill (potential/antipotential method)

The method of determining the firepower scores is a very difficult problem. There are several methods of computing firepower score values, such as military judgment and experience (RAND's ground force scoring system) (Allen P., 1992), historical combat performance derived from WWII and the Korean War, and results from high resolution simulations (e.g., Anti-Potential-Potential Method) (Parry & Hatman, 1992).

In Gozel's research (Gozel, 2000), the *ATLAS* ground attrition equations (Taylor, Yildirim, & Murphy, Hierarchy-of-Models Approach for Aggregated-Force Attrition,",,, 2000), RAND's Situational Force Scoring (SFS) (P. ALLEN, 1997) and Dupuy's Quantified Judgment Models (QJM) (Dupuy, 1987) are applied to three data sets that are extracted from the data on the Battle of Kursk.

Force Ratio Approaches - Correlation of Forces

Correlation of Forces is used primarily in planning, not in combat models. It was used in Soviet-style operations planning to allocate forces to a subordinate for a combat mission. A similar approach was used in U.S. operations planning to estimate whether various parts of a plan are feasible. This method could be used in a model where attrition is not the focus.

The Correlation of Forces Method (COFM) is a simply implemented, quantitative process for estimating the likelihood of attack success, assessing projected resource allocations, and monitoring current operations, and is applicable to the operational level of war and above. It is used in qualitative methods developed by the U.S. Army for

- Military Decision Making Process (MDMP)
- Battle Damage Assessment Process
- Course of Action (COA) Analysis
- War games

The COFM methodology has a long history. ((Hines, Petersen, & III, Soviet Military Theory from 1945-2000, 1983)) See (Hines, The Soviet Correlation of Forces Method, 1990), for a good summary of the background and origins of COFM.

Force Ratio Attrition Models - U.S. Army Capability Analysis

For Army Capabilities Analysis a base-type unit is selected and given the combat value 1.0. For instance a Striker vehicle might be chosen and assigned the value one. Units are assigned subjective values based on their perceived strength compared with base unit type, considering

- Primary type of equipment in unit (e.g., M1A1 versus M1 versus M60A3),
- Percent strength authorized, and
- Modifications subjectively based on intel estimate and current status of own unit.

The ratio of friendly to enemy estimated combat power is used to infer the capabilities of the friendly force. Example: U.S. Division planning for an operation against a threat division using Soviet-style equipment and organization.

Soviet-style COFM relies on force strength calculated using Standardized Unit Armament (SUA) scores for each weapon system. One might also use the comparable U.S. Army Weapons Effectiveness Index Weapons Unit Value (WEI/WUV) scores. The scores for each weapon system could also be taken from user input, or they could be based upon the "importance" score for each weapon as derived using an attrition methodology, such as ATCAL (Attrition Calibration) from a standard battle for the attack. The sum of the scores for all the attacker's weapon systems that are allocated to the attack becomes the attacker's force strength. This strength includes allocated reserves, indirect fire support, and interdiction assets. The defender's force strength is computed from the weapons systems allocated by the defender. COFM values are dynamic and can change due to losses in attacking or defending forces, or if unallocated reserves are assigned to a unit.

Force Ratio Attrition Models - U.S. Army Capability Analysis Example

U.S. Comparison Values (with respect to a BTR Battalion)

Maneuver:

- M113 Bn = 1.50
- M2 Bn = 2.00
- M1A1 Bn = 3.15
- M1 Bn = 3.00
- M60A3 Bn = 2.25
- ACR Sqdn = 2.75
- Div Cav Sqdn = 1.50
- Div Cav Sqdn(h) = 2.00
- Atk Hel Bn (AH64) = 4.00
- Atk Hel Bn (AH1) = 3.00

Artillery:
- MLRS Battery = 2.00
- 155mm SP Bn = 2.00

NOTE 1: All units are given an estimated value with respect to a BTR Battalion

NOTE 2: The unit commander and operations officer are advised to develop their own table to allow them to take into consideration local knowledge about the situation. This table does not contain real planning data, but only notional data. (Example extracted from US Army CGSC ST 100-9.)

Table X-X. Planning Rules of Thumb for Force Capabilities:

Friendly Mission	Friendly : Enemy	Notes
Delay	1 : 6	
Defend	1 : 3	Friendly deliberate defense
Defend	1 : 2.5	Friendly hasty defense
Attack	2.5 : 1	Enemy hasty defense
Attack	3 : 1	Enemy hasty defense
Counterattack	1 : 1	Attack enemy's flank

Force Ratio Attrition Models - U.S. Army Capability Analysis Example Solution

53 Mech Division				27 Gds Motorized Rifle Division			
Maneuver:				Maneuver:			
TYPE	BN	VALUE	TOTAL	TYPE	BN	VALUE	TOTAL
M113	4	1.5	6.0	BTR	6	x 1.0 =	6.0
M2	1	2.0	2.0	BMP	3	x 1.5 =	4.5
M1	2	3.0	6.0	T64(MRR)	3	x 1.8 =	5.4
M60	3	2.25	6.8	T64 (TR)	3	x 1.4 =	4.2
Cav	1	1.5	1.5	T64 (ITB)	1	x 2.0 =	2.0
Atk Hel	1	3.0	3.0	Div Recon	1	x 1.6 =	1.6
TOTAL			25.3	AT Bn	1	x 1.0 =	1.0
X %Strength			X .9	Atk Hel	1	x 3.0 =	3.0
Relative Cbt Power			22.8	TOTAL		=	27.7
				X % Strength		X	.8
Ratio for Maneuver: 22.8 : 22.2 = 1:1				Relative Cbt Power		=	22.2
				Artillery:			
Artillery:				TYPE	BN	VALUE	TOTAL
TYPE	BN	VALUE	TOTAL	122(SP)	2	x 2.0 =	4.0
155(SP)	3	2.0	6.0	122(T)	4	x 2.0 =	8.0
MLRS	1	2.0	2.0	152(SP)	1	x 2.0 =	2.0
TOTAL			8.0	MRL Btry	3	x 1.0 =	3.0
X %Strength			X .9	TOTAL		=	17.0
Relative Cbt Power			7.2	X % Strength		X	.8
				Relative Cbt Power			
Ratio for Artillery: 7.2 : 13.6 = 1:1.9							

ATLAS Model

The *ATLAS* theater level simulation uses a straightforward force ratio method. The simplicity of its structure is one of the main attractions of the *ATLAS* model. *TACWAR* is one of the simulations that use the ATLAS equations. In the combat attrition process of the *ATLAS* model, the casualty rates are determined by using simple equations for the attacker and the defender. The original casualty rates used in the *ATLAS* model were derived from data on 37 division level engagements in World War II and Korea (Parry & Hatman, 1992). Since the specific engagements are not documented, it is unknown as to whether the division-level data includes the battle of Kursk—though it is believed not to. If it is included, then the comparisons are not strictly independent. However, since Kursk would be the only one of a large number of engagements (37) the dependence will be very small. There is no published study on the validation of *ATLAS* equations for real combat data.

- The *ATLAS* theater level simulation uses a straightforward force ratio method.
- The simplicity of its structure is one of the main attractions of the *ATLAS* model.
- *TACWAR* is one of the simulations that use the *ATLAS* equations.
- In the combat attrition process of the *ATLAS* model, the casualty rates are determined by using simple equations for the attacker and the defender.
- The original casualty rates used in the *ATLAS* model were derived from data on 37 division level engagements in World War II and Korea (PARRY, 1992).
- There is no published study on the validation of *ATLAS* equations for real combat data.

Attrition Coefficient Calculation - Potential / Anti-potential ("Eigenvalue") Method

A different approach developed in the early 1970s determined the score for a weapon by observing what it could accomplish in a battle. In particular, the score for a weapon was defined as proportional to the total of the scores for all the enemy systems it kills. This definition leads to a circular system of eigenvalue equations that can be solved for the weapon score values (reference 4). The resulting scores are highly situation dependent and must be evaluated in the context of a battle scenario. The computational procedure is called the Potential Anti-Potential Method.

The Potential Anti-Potential method (or eigenvalue method) for computing weapon scores is significantly different from all the earlier methods. The earlier score computation formulas all depended entirely on the characteristics of the weapon itself to yield a score value that was (hopefully) useful independent of the enemy being faced and of the particular scenario. This goal was too ambitious, and thus there was a continual effort to change and improve the score definitions. The eigenvalue method depends on how the weapon capabilities interact with enemy vulnerabilities in a particular combat scenario. The computations include elements of the heterogeneous approach to aggregation, but eventually yield scores, indices, and force ratios for a homogeneous representation of unit combat power.

- A way to assign firepower scores.
- Avoids the problem of assigning scores based only on weapon's own characteristics, not on opposing targets' characteristics.
- Avoids some problems of subjectivity.
- CONCEPT: Let the value of a weapon system be directly proportional to the rate at which it destroys the value of enemy weapon systems.
- NOTE: Problem reduces to system of simultaneous linear equations.

The Potential Anti-Potential method for computing weapon system scores is defined by the following basic principle:

The value (score) of a weapon system is directly proportional to the rate at which it destroys the value of opposing enemy weapon systems.

Thus, the value of a system depends on its kill rates and on the value of the enemy systems it kills. Conversely, the enemy system values depend on the values of the friendly systems that they kill. Thus, the value definitions are circular.

Consider two opposing forces (called X and Y) made up of heterogeneous weapon systems. Suppose that the X force contains m different weapon system types and that the Y force contains n types. Let

X_i = the number of weapons of type i in the X force for $i = 1, 2, \ldots, m$,

and let

Y_j = the number of weapons of type j in the Y force for $j = 1, 2, \ldots, n$.

Define the weapon values (or scores) to be

SX_i = the value of one type i weapon in the X force, and

SY_j = the value of one type j weapon in the Y force.

Finally, define the kill rates:

K_{ij} = the rate at which one X_i system kills Y_j systems, and

L_{ji} = the rate at which one Y_j system kills X_i systems, for all possible combinations of the weapon indices $i = 1, 2, \ldots, m$ and $j = 1, 2, \ldots, n$.

So, $FPI_X = ⊡ \sum_i SX_i\, X_i$ and $FPI_Y = ⊡ \sum_j SY_j\, Y_j$

Attrition Coefficient Calculation - Potential / Anti-potential Method

The kill rates are assumed to have known non-negative numeric values, and we will solve for the score values. Kill rate values can be obtained from the killer-victim scoreboard output from a high-resolution simulation model. Note that they will implicitly depend on scenario details such as the composition of both forces, the force missions, target acquisition conditions, target selection rules, and the outcomes of one-on-one engagements.

The above concept and definitions lead to a system of simultaneous equations:

$$\text{for all } j,\ c_Y ⊡⊡ SY_j = \sum_i L_{ji} * SX_i \quad (1)$$

$$\text{for all } i,\ c_X ⊡⊡ SX_i = \sum_j K_{ij} * SY_j \quad (2)$$

In terms of these definitions, the basic valuation principle can be written as a system of equations, (1) and (2) above, where C_X and C_Y are the proportionality constants for the two forces.

This is a system of $m + n$ equations in $m + n + 2$ unknowns, since the two proportionality constants c_X and c_Y are not specified.

In matrix notation:

$$(3)\ c_Y\, ⊡\mathbf{SY} = \mathbf{L\, SX}$$

$$(4)\ c_X\, ⊡\mathbf{SX} = \mathbf{K\, SY}$$

Combining the equations gives a system of $m + n$ linear equations in the $m + n$ unknowns SX_i and SY_j for any given values of the proportionality constants. However, we will allow the mathematics to determine the values of C_X and C_Y also because then we can guarantee a solution in which all the scores are non-negative. The value equations can be expressed more compactly in matrix notation, (3) and (4) above.

To solve the value equations, substitute the expression for **SY** from equation (4) into equation (3) yielding

$$C_X\, C_Y\, \mathbf{SX} = \mathbf{K\,L\,SX}.$$

Similarly substituting for SX in equation (3) yields

$$C_X\, C_Y\, \mathbf{SY} = \mathbf{L\,K\,SY}.$$

If we define $\lambda = C_X \cdot C_Y$, then the above become

$$\lambda\, \mathbf{SX} = (\mathbf{K\,L})\, \mathbf{SX}, (5)$$

$$\lambda\, \mathbf{SY} = (\mathbf{L\,K})\, \mathbf{SY}, (6)$$

which can be recognized as a pair of eigenvalue problems for the non-negative matrices $\mathbf{K} \cdot \mathbf{L}$ ($m \times m$) and $\mathbf{L} \cdot \mathbf{K}$ ($n \times n$). The eigenvalue is λ and the eigenvectors are **SX** and **SY**.

Although we cannot present the details here, the Frobenius Theorem guarantees that there exists a real, non-negative, largest eigenvalue E (the same for both equation systems), and there exist non-negative eigenvectors **SX** and **SY** (unique up to a scale factor) that satisfy the equations of the eigenvalue problem.

The resulting solutions are consistent with the original basic principle for valuing weapon systems, so they can be used for score values. Using the scores SX_i and SY_j we can compute aggregated unit index values and force ratios in the ordinary fashion.

SOLUTION PROCESS for 2×2 matrix:

1. Solve for λ:

Characteristic polynomial is:

$$det(X - \lambda I) = 0$$

$$\begin{vmatrix} C_{11} - \lambda & C_{12} \\ C_{21} & C_{22} - \lambda \end{vmatrix} = \lambda^2 - (C_{11} + C_{22})\lambda + (C_{11}C_{22} - C_{12}C_{21})$$

Let $d_1 = C_{11} + C_{22}$ and $d_0 = C_{11}C_{22} - C_{12}C_{21}$, then

2. Substitute for λ in equations (3) and (4) and solve for X_{ij}.

$$\lambda = \frac{1}{2}\left(d_1 \pm \sqrt{d_1^2 - 4d_0}\right)$$

<u>GIVEN</u>:

X force of 2 systems, strengths $X_1 = 200$, $X_2 = 150$

Y force of 2 systems, strengths $Y_1 = 75$, $Y_2 = 100$

X_1 kills Y_1 at rate .03, Y_1 kills X_1 at rate .04

X_1 kills Y_2 at rate .05, Y_1 kills X_2 at rate .02

X_2 kills Y_1 at rate .03, Y_2 kills X_1 at rate .04

X_2 kills Y_2 at rate .02, Y_2 kills X_2 at rate .01

So

$$\mathbf{K} = \begin{bmatrix} 0.03 & 0.05 \\ 0.03 & 0.02 \end{bmatrix}, \quad \mathbf{L} = \begin{bmatrix} 0.04 & 0.02 \\ 0.04 & 0.01 \end{bmatrix}$$

$$\mathbf{KL} = \begin{bmatrix} 0.0032 & 0.0011 \\ 0.0020 & 0.0008 \end{bmatrix}, \quad \mathbf{LK} = \begin{bmatrix} 0.0018 & 0.0024 \\ 0.0015 & 0.0022 \end{bmatrix}$$

for $\lambda = c_X c_Y$, solve for λ in $\lambda \mathbf{SX} = \mathbf{K L SX}$.

First, solutions **SX** and **SY** can easily be computed by standard eigenvalue programs. Unfortunately, the solutions are not unique. By examining the original equations

$$C_X \mathbf{SX} = \mathbf{K SY} \text{ and } C_Y \mathbf{SY} = \mathbf{L SX},$$

we can see that if **SX** and **SY** solve the equations then so will the scalar multiples $MX^*\mathbf{SX}$ and $MY^*\mathbf{SY}$ for any arbitrary scale factor multipliers MX and MY. In the new solution, the values of the proportionality constants C_X

and C_Y will adjust to absorb the scale changes. If a solution vector **SX** is multiplied by the scalar MX, then the relative value between two different X force weapons will remain unchanged:

$$SX_1/SX_2 = (MX * SX_1)/(MX * SX_2).$$

Thus the scale factors do not affect weapon comparisons within the same force. However, weapon comparisons between the X force and the Y force are clearly changed when the scores are scaled:

$$SX_1/SY_2 = (MX * SX_1)/(MX * SY_2).$$

If $MX \neq MY$. The same result is true for the aggregate force ratio; scaling the scores changes the force ratio value by a factor of MX/MY. Thus the method chosen to scale the score vectors is extremely important. Several scaling methods have been proposed, but we will concentrate on the scaling method used in the IDAGAM attrition structure.

Select some major weapon system from the X force, one that will engage numerous Y force system types (say we choose X system 1). Scale the X force score vector so that the new $NSX_1 = 1.0$ thus determining new values for all the other NSX_i. This can be accomplished by setting the scale factor to

$$MX = 1.0/SX_1.$$

$d_1 = 0.0032 + 0.0008 = 0.004$, and

$d_0 = (0.0032 * 0.0008) - (0.0011 * 0.0020) = 0.00000036$,

$\lambda = 0.5(0.004 \pm \sqrt{(0.0042 - 4 * 0.00000036)}\,)$,

$\lambda^* = max(\lambda_1, \lambda_2) = 0.003907878.$

Let $SX_1 = 1.0$

Let $C_X = C_Y = \sqrt{\lambda}$

Since λ **SX** = **K L SX**, then

(**K L** $- \lambda$ **I**)**SX** = **0**.

$$\left(\begin{bmatrix} 0.0032 & 0.0011 \\ 0.0020 & 0.0008 \end{bmatrix} - \begin{bmatrix} 0.003907878 & 0.0 \\ 0.0 & 0.003907878 \end{bmatrix}\right) \begin{bmatrix} SX_1 \\ SX_2 \end{bmatrix} = \begin{bmatrix} 0 \\ 0 \end{bmatrix}$$

Since $SX_1 = 1.0$, we can solve for SX_2 in two ways:

$$-0.000707878 SX_1 + 0.0011 SX_2 = 0 \text{ or}$$

$$0.002 SX_1 - 0.003107878 SX_2 = 0$$

Either way gives $SX_2 = 0.64353$

Now solve for SY_1 and SY_2 using eqn (1): c_Y **SY** = **L SX**

$$\sqrt{\lambda} \begin{bmatrix} SY_1 \\ SY_2 \end{bmatrix} = \begin{bmatrix} 0.04 & 0.02 \\ 0.04 & 0.01 \end{bmatrix} \begin{bmatrix} 1.0000 \\ 0.6435 \end{bmatrix}$$

where $\sqrt{\lambda} = \sqrt{0.003907878} = 0.062513$.

Thus,

$$\begin{bmatrix} SY_1 \\ SY_2 \end{bmatrix} = \frac{1}{0.062513} \begin{bmatrix} 0.0529 \\ 0.0464 \end{bmatrix} = \begin{bmatrix} 0.8458 \\ 0.7428 \end{bmatrix}$$

Now compute the Fire Power Index for each force, FPI_X and FPI_Y, and then FR:

The force ratio, FR is given by

$$FR = \frac{FPI_X}{FPI_Y} = \frac{(200 \times 1.0) + (150 \times 0.6435)}{(75 \times 0.8458) + (100 \times 0.7428)} = \frac{433.091}{137.712} = 2.153$$

Now to assess actual casualties lost by the X and Y forces, enter a table indexed by FR (and perhaps other factors) to find a percent of force lost. Apply this percentage to each weapon type.

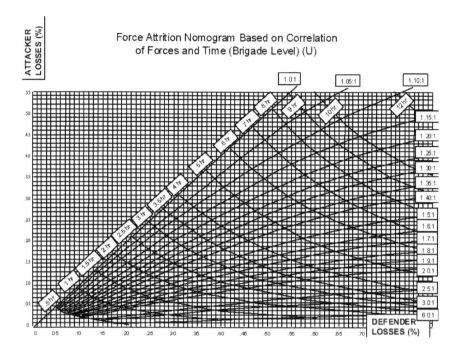

The score values that result from the eigenvalue solutions are scenario dependent because they depend on the kill rates A_{ji} and B_{ij}. The circular nature of the value equations makes the relationships among the scores complex. The kill rates also depend implicitly on the number of weapon systems in the forces since that influences target engagement opportunities.

Typically, any change in any of the kill rates will cause all of the score values for both sides to change in ways that are hard to predict. Thus we should not consider the eigenvalue scores to be a measure of long term inherent value of a weapon system, but rather only of transient value in a specific situation. Indeed, the Institute for Defense Analysis Gaming Model (IDAGAM) theater combat simulation, which was the first simulation to use eigenvalue scores, reevaluates the score values for each day of combat.

Note: IDAGAM is a deterministic theater level model of ground and air combat developed in 1974 by the Institute for Defense Analysis. The basic maneuver unit is generally a division or an independent brigade.

Attrition in DoD Aggregated Models

- TACWAR – Force Ratio (Anderson)

- Vector In Command – Lanchester-type (Bonder-Farrell)
- WARSIM – Lanchester-type (Bonder-Farrell)
- Concepts Evaluation Model (CEM) – Lanchester-type (Clark)

Other Aggregated Models

Joshua Epstein (who developed the "piston model") found Lanchester equations too limiting to describe complex combat situation. In "The Calculus of Conventional War; Dynamic Analysis Without Lanchester Theory", Epstein contends that Lanchester equations fail to capture warfare's basic dynamics and present a fundamentally misleading picture of war. He then presents new, alternative equations of his own (Epstein J. M., 1985). These, he contends, more accurately represent the core dynamics to which Lanchester theory is oblivious.

$$W(t) = W(t-1) + \left(\frac{W_{max} - W(t-1)}{1 - \alpha_{dT}} \right) (\alpha_d(t-1) - \alpha_{dT})$$

$$\alpha_g(t) = \alpha_g(t-1) - \left(\frac{\alpha_{aT} - \alpha_g(t-1)}{\alpha_{aT}} \right) (\alpha_a(t-1) - \alpha_{aT})$$

Trevor N. Dupuy (Col, USA, Ret) developed the Quantified Judgment Model (QJM) to use historical data to evaluate and predict the outcome of armed conflict (Dupuy, 1987).

Behavioral emulation simulates the movements of units and observes their interactions and the resulting outcomes. This emulation can be done with real troops in an exercise, with models on a sand table, or most recently with software agents in a simulated world. Entity-based models such as OOS and Combat XXI assign a single agent to each entity, following the standard MAS agenda. A new modeling construct, the polyagent, takes this trend one step further, and uses several agents to model each construct.

An aggregate state space model is an equation relating the change in the sides' strengths (attrition) to various state variables and the duration of the time step, using the initial state-variable values of the time step and treating the combatants as all "in" the same location. This aggregate result may or may not correspond to a Lanchester square law, linear law, or something similar.

Chapter 11. Behavioral Modeling

Because of its complexity, behavioral modeling has traditionally been very basic. The goal has been to provide military vehicles and units with the ability to react to basic events in the absence of human intervention. These models allowed aircraft on patrol to "decide" to return to base when getting low on fuel, rather than continuing until the aircraft falls to the ground. Ground units respond to enemy attacks by focusing firepower on the aggressor rather than blindly continuing their preprogrammed mission. Algorithms like these have been the extent of behavioral modeling for many years. However, more recent models have attempted to provide more reasoning capabilities to simulated objects. Most notable among these systems have been the Semi-Automated Forces (SAF) or Computer Generated Forces (CGF) systems that are used to stimulate virtual training audiences. These allow one operator to play the part of many vehicles or several platoons with the aid of embedded behavioral models.

Behavioral Representation

The approach taken by most of these models is to replicate the product of human decision making, rather than the process. Since we do not completely understand the inner workings of the human mind, it is much more feasible to gather information about human reaction to certain situations than it is to represent the process of thinking about that information. However, research in the area of intelligent agents is leading to models of independent, emergent behavior derived from the interactions of multiple stimuli on an object. Current systems make use of the following technologies from the artificial intelligence field to model human decision making: finite state machines, means-ends analysis, constraint satisfaction, expert systems, knowledge based systems, and traditional planning. Evaluations have been done on the applicability of Petri nets, Markov chains, case based reasoning, fuzzy logic, neural networks, genetic algorithms, and adaptive behavior. Each of these techniques has strengths and weaknesses for military decision making. Researchers familiar with both the simulation and AI fields are developing techniques specifically designed for this problem.

Artificial Intelligence

The original promise of artificial intelligence (AI) was that it would provide a flexible framework for the development of intelligent systems capable of extracting information, making decisions, and applying control. Initial work in AI concentrated on universal aspects of this problem with limited success. It soon became clear that the spaces through which one had to search in order to choose optimal actions were vast, much larger than computing resources could handle by exhaustive search techniques. It became important to find heuristics that limited the search to a tractable subspace, which would contain a good choice.

Behavior Categories

There is a need for models of several types of human activity. Individual skills represent the activity of a single human performing a defined skill like movement, reaction to external events, and using tools. Behavioral modeling in military simulations usually includes aspects of human military decision making.

Figure 11-1. Agent behavior categories

Until recently, the command and control processes and especially the decision process were - if at all - hard-coded in the closed combat simulation models. Although in other application areas expert systems and rule systems as well as neural nets were used to find solutions to similar problems, in the closed combat community these approaches are still not used sufficiently.

Military Decision Making (METT-TC)

- Mission – Mission and tasks to perform
- Enemy – Enemy units to handle
- Terrain – Types of terrain to reason on
- Troops – Friendly troops to employ
- Time – Time available to perform planning
- Civilians – Consider impact on civilians

The problem with hard-coded decisions is always, that they are evaluated only for a special environment. Being transferred to another environment the decision model is seldom capable to meet the new requirements. Self-optimization and the skill to adapt to new situations are nowadays more often required by such systems.

In constructive and virtual simulations, the doctrinal approach of METT-TC is traditionally implemented as either human-in-the-loop, or with simple structures such as decision trees or scripting.

Agent Types

An **agent** is someone with expertise who is entrusted to go out and act on your behalf. There are three basic dimensions for agents are that they should...

1. Be autonomous
2. Cooperative
3. Learn

Agents are categorized based upon the dimensions that are modeled.

- Collaborative agents are autonomous and cooperate.
- Interface agents are autonomous and learn
- Collaborative agents cooperate and learn
- Intelligent agents demonstrate some notion of all three dimensions.

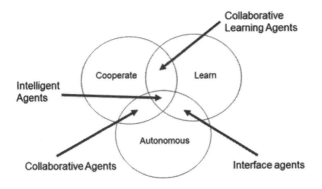

Figure 11-2. Types of agents and their domains

Basic Intelligent Agent

"We want to build intelligent actors, not just intelligent thinkers. Indeed, it is not even clear how one could assess intelligence in a system that never acted -- or, put otherwise, how a system could exhibit intelligence in the absence of action."

- (Pollack, 2006).

An agent can be defined as, *"Anything that can be viewed as perceiving its environment through sensors and acting upon that environment through effectors"* (Russell & Norvig, 1995). Alternatively, an agent is, *"A self-contained software element responsible for executing part of a programmatic process, usually in a distributed environment"* (Guilfoyle & Warne, 1994).

Crucial notions of agents include:
- Autonomous
- Personalizable
- Reactive
- Risk and trust
- Domain
- Graceful degradation
- Cooperation
- Social
- Self-directed

Virtual agents feature computerized characters that look, sound, move and seemingly think like real people. Making computers human is an idea as old as computers themselves, and what was initially a wild science fiction fantasy is gradually becoming workable. From the chilling 2001: A Space Odyssey's HAL 9000 to robotic newsreader Ananova, virtual creatures have become part of our collective culture. The potential of computerized agents or entities that are autonomous, self-directed, reactive and social— just like humans—can be estimated only in the realm of the imagination. Already, such agents have been built to present the weather on mobile phones, drive trucks, monitor environments designed to support life on other planets and perform many other sophisticated tasks. Computers are good at doing what they are told, but in this field, they are required to reach their own conclusions. The complex computer code beneath their 'skins' is designed to make them react to situations as real people do— unpredictably.

Agents involve some notion of:

- **trust**: the agent will do what you think it will do,
- **personalizability**: the agent can be either learn or be explicitly taught what to do for each individual user, and
- **autonomy**: the agent is allowed to take at least some actions on the user's behalf, without permission or perhaps even notification.

Agent Categories

Agents are generally categorized as reactive or deliberative. A reactive agent makes a decision in response to each input. It is not goal-oriented, but is much more of a data processing machine that performs certain actions in response to individual input.

Deliberative agents maintain a goal and respond to the environment according to that goal. Therefore, many inputs may be received before a single action or decision is made. Although an older (traditional) AI approach, these agents may be most appropriate for the complexities of command decisions. A hybrid of these two approaches may provide a better agent model for future simulations (Arkin, 1998).

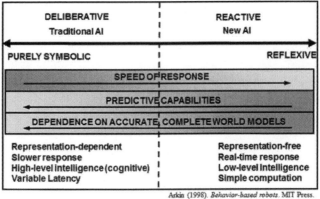

Figure 11-3. Two types of agents: deliberate and reactive

Reactive Agent Architecture

Agents are an encapsulating technique for organizing a behavioral model's interaction with and effects on the outside world. An agent architecture may encapsulate any behavioral model. It is defined by the fact that it is distinctly separate from its environment; uses sensors to receive input about the environment; and employs effectors to interact with that environment.

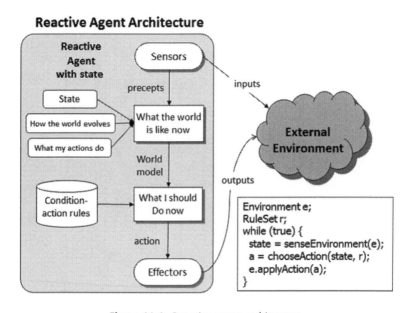

Figure 11-4. Reactive agent architecture

194

The reactive agent architecture depicted above indicates that sensors receive data from the external environment, uses its precepts to perceive the environment (determine what the world is like now), employs a world model to select action based on a set of condition-action rules, implements the actions, and through its effectors, translates this action to the external environment.

Adding states to the model enhance the agent so that it can perform a sort of "what-if" analysis, i.e., what will various actions do when the external world is in state A as it evolves to the next state.

Deliberate Agent Architecture

In the deliberative agent model, we add an additional step that asks "what will the world be like if I do action A?", while considering transition states resulting from how particular actions may evolve the world. This depicts a goal-based agent model, which maintains a perception of its environment, based on complex reasoning.

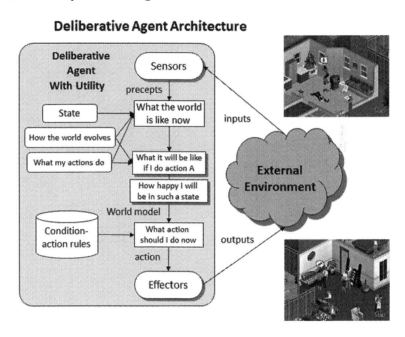

Figure 11-5. Deliberate agent architecture

Adding the agent considers how happy it will be in such a state, say state A, we evolve to a goal-based to a utility-based agent model. These models add a degree of "happiness" into their decision making process. When there are conflicting goals, the utility function specifies the appropriate trade-off. Moreover, when there are several goals that the agent can aim for, none of which can be achieved with certainty, utility provides a way in which the likelihood of success can be weighted up against the importance of the goals and how the agent interaction depends on them.

Anatomy of a Cognitive Agent

This chart illustrates one possible architecture for realizing a cognitive computational system (a blend of reactive and deliberative agent models). This diagram shows the relationships and connectivity among the three major processes that are usually associated with cognition. In addition, it shows the relationship between these processes and the machine's perception (sensors), action (effectors), and the environment. A cognitive agent may also be called an intelligent agent or smart agent (Brachman & Levesque, 2004).

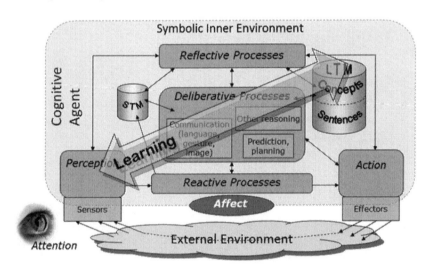

Figure 11-6. Anatomy of a cognitive agent

Agent Reasoning Mechanisms

Agent structures may be used to encapsulate any of the traditional AI techniques for solving a problem. Agents are not another AI technique, but rather a useful method for organizing reasoning behavior within an environment.

Finite State Machines

Finite State Machines (FSM), also known as Finite State Automation (FSA), at their simplest, are models of the behaviors of a system or a complex object, with a limited number of defined conditions or modes, where mode transitions change with circumstance.

Finite state machines consist of 4 main elements:

- states which define behavior and may produce actions
- state transitions which are movement from one state to another
- rules or conditions which must be met to allow a state transition
- input events which are either externally or internally generated, which may possibly trigger rules and lead to state transitions

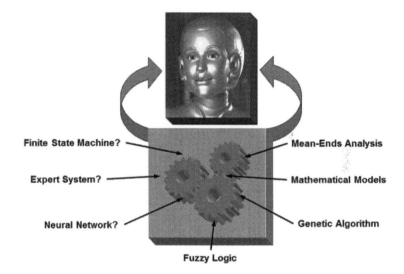

Figure 11-7. Different behavior modeling methods

A finite state machine must have an initial state which provides a starting point, and a current state which remembers the product of the last state transition. Received input events act as triggers, which cause an evaluation of some kind of the rules that govern the transitions from the current state to other states. The best way to visualize a FSM is to think of it as a flow chart or a directed graph of states, though as will be shown; there are more accurate abstract modeling techniques that can be used.

The figure below shows the states of a Rocket from the game *Quake*. This is representation is a subjective interpretation of code. A rocket in *Quake* is a projectile fired from the Rocket Launcher weapon/item which may be possessed and operated by a human player. The light grey boxes are the states, the darker gray the triggers and the arrows are the state transitions. The black boxes show the entry point and exit points of the system. The diagram shows the full life cycle of the rocket projectile within the game. It is interesting to note that the projectile is spawned into existence as the product of an action of another FSM, namely that of the "rocket launcher" from its "fire" action. When the projectile instance dies it is removed from the game, and no longer exists.

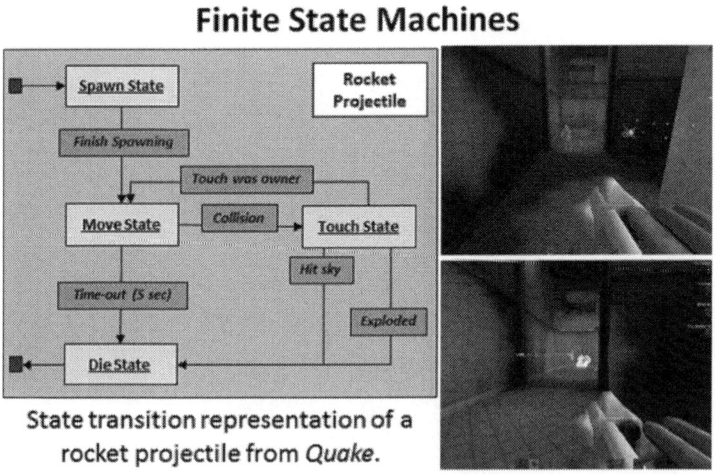

Finite State Machines

State transition representation of a rocket projectile from *Quake*.

Figure 11-8. *Quake* uses a finite state machines for various behaviors

FSM is a technique that originated in mathematics. Initially, they were used to represent spoken languages diagrammatically. However, they have been used in games since the dawn of AI in video games in the 1970s. Pong, Pac-man and Space Invaders all made use of FSMs and modern games continue to use them today.

However, FSMs can become limiting because entities that use FSMs only have as many states as the programmer gives them. As a result, FSMs can show little unpredictability – if they see an enemy shooting, the AI returns fire; if the AI sees another enemy, it does exactly the same thing. The popular game *Half-Life* attempted to surmount this problem in an interesting way. The *Half Life* programmers developed 'schedule-driven state machines'. These provide the AI player with multiple different 'schedules' of actions for any given state, adding another layer of complexity to the state machine. For example, if the AI soldier saw a human player, it would first look at other factors such as its health or ammunition status to enable it to decide which schedule to execute. In one case, it may stand and fight whereas in others it may retreat, adding to the realism of the entity's behavior. *Half Life* was undoubtedly a step forward for game AI, it showed what could be accomplished with even the most basic AI techniques.

Mathematical Models

Humans perform many operations that are based on measurable environmental conditions and a reasoning process that can be captured in a mathematical model. These operations require the human to operate as a calculator or data integrator and are easily replicated mathematically. In this context mathematics includes probability and statistics whish are often used for stochastic decision making.

$$delivery_{angle} = F\left(weapon_{angle}, weapon_{count_{weapon_{type}}}\right)$$

$$delivery_{angle} = \frac{\sum_{i=1}^{N}(weapon\_angle_i * weapon\_count_i)}{\sum_{i=1}^{N} weapon\_count_i}$$

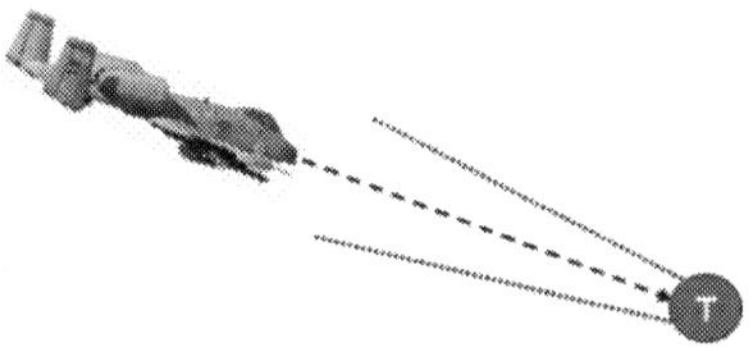

Figure 11-9. Direct fire air attack on a target

Markov Chains

A Markov chain is an adaptation of a FSM where the transitions among states are probabilistic in nature. Instead of transitioning from one specific state to another in a deterministic manner, variability is added through a stochastic method of determining the next state.

As in the classical definition, a Markov chain can be represented as a graph of nodes that are states and edges or links that represent actions that lead to new states. In addition, there are transition probabilities, and the notion that, for any specific level of a hierarchy, any given state is only conditional on immediately previous states.

- Stochastic decision state transition engine
- States linked by probability spaces
- $P[A|B]$ is probability of A given B
- Initial state – random number – new state

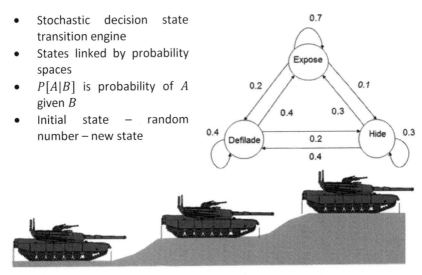

Figure 11-10. A Markov chain exposure/detection mode a tank in different fighting positions

Fuzzy Logic

While the Department of Defense has funded the largest computer wargames ever created (WARSIM, JWARS, TACOPS) they are all curiously devoid of strategic or even tactical AI. For example, WARSIM does not use AI, since Humans play the enemy. However, it has some smart behaviors as well as a Fuzzy Logic system that attempts to portray some of the soft factors of war.

Fuzzy Logic shows some promise for the creation of Human-Level AI. Humans rarely think in absolute terms. The driver's education book states definitively that you are to signal a turn precisely100 ft. before the intersection but people never get out and measure the exact distance.

The same is true with military or strategic axioms. It is "good" to get behind an enemy. It is good to cut an enemy's lines of supply and communication. However, there are no precise mathematical definitions for the terms.

In 1985 fuzzy logic was used in the commercial wargame, *UMS: The Universal Military Simulator*. The program "understood" concepts such as "line" and "flanks". This was accomplished by first drawing an imaginary box around all the units in an army, then determining the army's facing by comparison with the enemy army. It was then easy to determine left and right flanks as well as the 'center' of the army's lines (a double envelopment is simply a simultaneous left and right flank attack).

Figure 11-11. Universal Military Simulation behavior is based on fuzzy logic

It is reasonable to assume that fuzzy logic will play a role in the development of Human-Level AI.

Fuzzy Rules

Human beings make decisions based on rules. Although, we may not be aware of it, all the decisions we make are all based on computer like if-then statements. If the weather is fine, then we may decide to go out. If the forecast says the weather will be bad today, but fine tomorrow, then we make a decision not to go today, and postpone it till tomorrow. Rules associate ideas and relate one event to another.

Fuzzy machines, which always tend to mimic the behavior of man, work the same way. However, the decision and the means of choosing that decision are replaced by fuzzy sets and the rules are replaced by fuzzy rules. Fuzzy rules also operate using a series of if-then statements. For instance, if X then A, if y then b, where A and B are all sets of X and Y.

Neural Networks

Surprisingly it is the commercial gaming industry that is showing the most interest in Neural Networks. I did not find any references to Neural Nets currently being employed for military applications. André LaMothe, a very respected writer about computer game programming, has produced an in-depth piece on the subject (LaMothe, Building Brains Into Your Games, 1999) (LaMothe, Neural Netware, 1999) (here neural.doc).

A neural network is an attempt to simulate the inner working of a biological neuron with software (Smith L. , 1996). Neural networks are attempting to model the inner workings of the human brain (Jain, Mao, & Mohiuddin, 1996). For the purposes of this paper, biological neurons will not be looked at because it is not relevant to the focus of this paper. For a detailed explanation of biological neural network see (Seaman, 2001).

- Mimic natural intelligence
    - Networks of simple neurons
    - Highly interconnected
    - Adjustable weights on connections
    - Learn rather than program
- Architecture is different
    - Brain is massively parallel
        - 1011 neurons
    - Neurons are slow
        - Fire 10-100 times a second
    - Brain is much faster
        - 1014 neuron firings per second for brain
        - 106 perceptron firings per second for computer

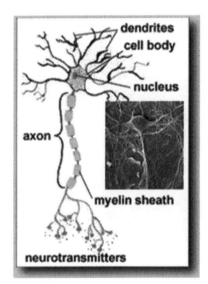

Figure 11-12. Characteristics of neural networks

Neural networks are made up of interconnected neurons. "Each neuron has a certain number of inputs, each of which have a weight assigned to them. The weights simply are an indication of how 'important' the incoming signal for that input is. The net value of the neuron is then calculated - the net is simply the weighted sum, the sum of all the inputs multiplied by their specific weight. Each neuron has its own unique threshold value, and if the net is greater than the threshold, the neuron fires (or outputs a 1), otherwise it stays quiet (outputs a 0). The output is then fed into all the neurons it is connected to" (Matthews, 2000).

Neural Networks in Fields of Battle

One of the most ambitious shareware wargames ever made, *Fields of Battle* (FoB) was a strategic simulation of World War I on the strategic level. While the avid wargamers would be disappointed at the absence of tactical options and relatively simplistic battle interface that is more reminiscent of parlor boardgames than tabletop wargames, Fields of Battle deserves recognition for introducing many interesting concepts to the genre, especially the first use of a neural network in a commercial computer game.

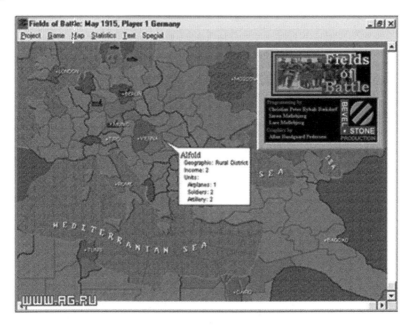

Figure 11-13. Fields of Battle screenshot

What made the game very interesting, though, is what went on beneath the facade of simplicity. The game used full-fledged neural networks for the computer player AI. This means, for example, that the computer supposedly learns from its mistakes, and adapts to the human players' strategies. This made FoB the first commercial game that that used neural networks of any kind (beating out Battlecruiser 3000 for that honor, despite claims to the contrary).

Battlecruiser: 3000 AD (BC3K) was a "your-ship-alone-vs-the universe" style of game in the tradition of Star Control. A neural network drove every NPC (non-player character) in the game. This included each of the 125-crew members of your ship, which was quite impressive technically. The computer opponents also used neural networks to guide negotiations, trading, and combat.

Genetic Algorithms (GA)

A genetic algorithm (GA) is a search heuristic that mimics the process of natural evolution. This heuristic is routinely used to generate useful solutions to optimization and search problems. Genetic algorithms belong to the larger class of evolutionary algorithms (EA), which generate solutions to optimization problems using techniques inspired by natural evolution, such as inheritance, mutation, selection, and crossover. They are a computational analogy of adaptive systems. They are modeled loosely on the principles of the evolution via natural selection, employing a population of individuals that undergo selection in the presence of variation-inducing operators such as mutation and recombination (crossover). A fitness function is used to evaluate individuals, and reproductive success varies with fitness.

Computer simulations of evolution started as early as in 1954 with the work of Nils Aall Barricelli, who was using the computer at the Institute for Advanced Study in Princeton, New Jersey (Barricelli, Esempi numerici di processi di evoluzione, 1954) (Barricelli, Symbiogenetic evolution processes realized by artificial methods, 1957). His 1954 publication was not widely noticed. Starting in 1957 (Fraser, Simulation of genetic systems by automatic digital computers. I. Introduction, 1957), the Australian quantitative geneticist Alex Fraser published a series of papers on simulation of artificial selection of organisms with multiple loci controlling a measurable trait. From these beginnings, computer simulation of evolution by biologists became more common in the early 1960s, and the methods were described in books by Alex Fraser and Donald Burnell (Fraser & Burnell, Computer Models in Genetics, 1970) and Jack Crosby

(Crosby, 1973). Fraser's simulations included all of the essential elements of modern genetic algorithms. In addition, Hans-Joachim Bremermann published a series of papers in the 1960s that also adopted a population of solution to optimization problems, undergoing recombination, mutation, and selection. Bremermann's research also included the elements of modern genetic algorithms. Other noteworthy early pioneers include Richard Friedberg, George Friedman, and Michael Conrad. Many early papers are reprinted by David Fogel (Fogel, 1998).

Although Barricelli, in work he reported in 1963, had simulated the evolution of ability to play a simple game (Barricelli, Numerical testing of evolution theories. Part II. Preliminary tests of performance, symbiogenesis and terrestrial life, 1963), artificial evolution became a widely recognized optimization method as a result of the work of Ingo Rechenberg and Hans-Paul Schwefel in the 1960s and early 1970s – Rechenberg's group was able to solve complex engineering problems through evolution strategies (Rechenberg, 1973) (Schwefel, Numerische Optimierung von Computor-Modellen mittels der Evolutionsstrategie : mit einer vergleichenden Einführung in die Hill-Climbing- und Zufallsstrategie, 1977) (Schwefel, Numerical optimization of computer models, 1981). Another approach was the evolutionary programming technique of Lawrence J. Fogel, which was proposed for generating artificial intelligence. Evolutionary programming originally used finite state machines for predicting environments, and used variation and selection to optimize the predictive logics. Genetic algorithms in particular became popular through the work of John Holland in the early 1970s, and particularly his book Adaptation in Natural and Artificial Systems (Holland, 1975). His work originated with studies of cellular automata, conducted by Holland and his students at the University of Michigan. Holland introduced a formalized framework for predicting the quality of the next generation, known as Holland's Schema Theorem. Research in GAs remained largely theoretical until the mid-1980s, when The First International Conference on Genetic Algorithms was held in Pittsburgh, Pennsylvania.

The Algorithms

1. Randomly generate an initial population $M(0)$.

2. Compute and save the fitness $u(m)$ for each individual m in the current population $M(t)$.

3. Define selection probabilities $p(m)$ for each individual m in $M(t)$ so that $p(m)$ is proportional to $u(m)$.

4. Generate $M(t + 1)$ by probabilistically selecting individuals from $M(t)$ to produce offspring via genetic operators.

5. Repeat step 2 until satisfying solution is obtained.

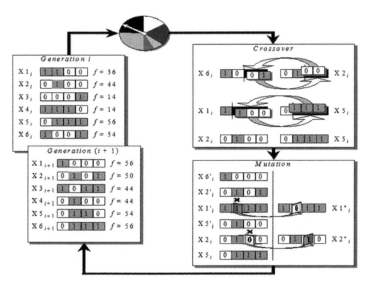

Figure 11-14. Typical Genetic Algorithm (GA)

Genetic Algorithms in bSerene

bSerene is a game that sports an AI which drives monsters that try to hunt down and kill your player. The monsters do not know where you are; they only gain information based on line-of-sight observations and communications with other monsters that might have seen or fought you. The way the monsters use such information can be used to manipulate their behavior, and a good player of this game will have to observe how the monsters react to his doings and learn to lead them into traps and split their groups.

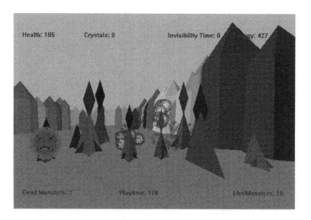

Figure 11-15. Screenshot from *bSerene*

AI aspects of the game, besides the monsters communicating on your location and learning it from one another, include an A-Life component in which the monsters can spawn baby monsters if sufficient numbers of them get together. Left to themselves the new monsters can quickly outmatch the player's ability to fight them, so the player is forced to move against them and try to break their spawning groups up.

The monsters themselves apparently use genetic algorithms between games to slowly evolve strategies of movement and search to engage the player. This makes the game somewhat like *Cloak, Dagger, and DNA* (CDDNA) which also uses GAs for similar purposes. The monsters will slowly adapt to the player's style of play over time, which can make for some formidable opponents.

Genetic Algorithms in Return Fire

Return Fire 2 (RF2) was the follow-up to a fairly popular action shooter game released in 1996. The new game promised larger maps, more vehicles, and a true 3D engine. What made this game interesting from an AI point of view is that it claimed to use genetic algorithms for the game AI. According to a preview article in the July, 1998 issue of Next Generation magazine (now, Edge-Online.com), the AI "...was designed by analyzing thousands of games, so enemies are finally capable of developing strategies worthy of human opponents." Another reviewer said, "The computer AI is very cunning and quite crafty. Even in the single-player mode *Return Fire 2* is a very intense action/strategy game" (Ganesh, 1998).

Figure 11-16. Screenshots of *Return Fire 2* (RF2)

Path Planning Mechanism using Genetic Algorithms

Porto, et al., demonstrated the feasibility of using genetic algorithms to solve platoon-level tactical problems. They developed a modified version of ModSAF (called MEWS) that focused on platoon-level course of action generation in an environment where two competing platoons must encounter each other on the way to their objectives. Different goal parameters can be set, such as the importance of timely arrival at the objective, the importance of survival, or the importance of eliminating enemy tanks. An adaptive algorithm drives the behavior of one or both sides in the conflict, and the genetic algorithm compares possible tactics based on the success parameters.

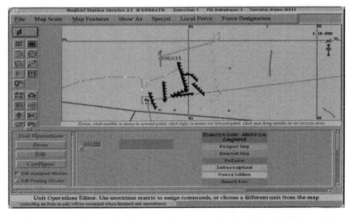

Figure 11-17. Path planning mechanism using genetic algorithms (Boissonnière, 1999)

Evolving Neural Networks (ENN)

The applications of EAs in Artificial Neural Networks (ANNs) are mostly concentrated in finding suitable network topologies and then training the network. EAs can quickly locate areas of high quality solutions when the domain is very large or complex. This is important in ANN design and training where the search space is infinite, highly dimensional and multimodal.

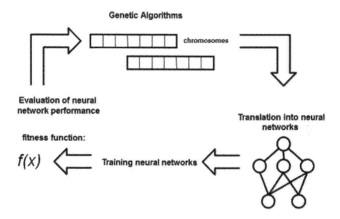

Figure 11-18. How an evolving neural network (ENN) works

The evolution of connection weights introduces an adaptive and global approach to training. Unlike gradient-based training methods, viz. back propagation, EAs rely on probabilistic search techniques and so, even though their search space is bigger, they can ensure that better solutions are being generated over generations. Optimal network architectures can also be evolved to fit a given task at hand. The representation and search operators used in EAs are two most important issues in the evolution of architectures. It has been shown that EAs relying on the crossover operator do not perform very well in searching for optimal network topologies.

ENN in MANA

Map Aware Non-uniform Automata (MANA) is an agent-based model developed by the Operations Analysis group at Defence Technology Agency in New Zealand to create a complex adaptive system to be used to examine combat (Lauren & Stephen, 2002). MANA has been used in a number of studies: modeling civil violence management, the modeling of maritime surveillance and coastal patrols, investigating modern warfare as a complex adaptive system and a range of studies carried out at the bi-annual Project Albert meetings.

MANA is an original piece of software that builds on earlier works such as ISAAC/EINSTEIN and the evolving Archimedes model. The version of MANA used is a beta, revised in April of 2001. Some of the features that differentiate it to other MAS models are situational awareness, a terrain map, waypoints, and event-driven personality changes

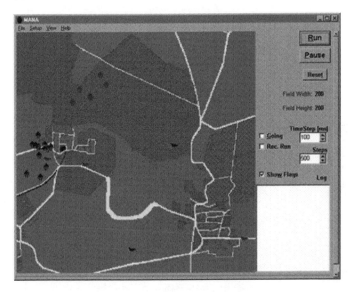

Figure 11-19. Screenshot from MANA

MANA's architecture allows for easy adaptation of the generalization of entities. Like our framework it has a sensor range, a firing or weapon range, movement range or speed. MANA also has a "firepower" setting represents a single shot kill probability or probability to hit as called in the framework. The durability of an agent can be modeled using MANA's "number of hits to kill" setting for each agent.

The lethality of a weapon does not have a direct relation to any setting in MANA, but there is a "max targets per step" setting that determines how many agents can be engaged at one time. This setting could be used as a type of weapon lethality, thus allowing one agent to engage more than one enemy.

There is also not communication range setting in MANA. Agents are able to communicate enemy in their sensing range to the HQ agent/squad, regardless of their distance to that HQ agent/squad. Conversely, the HQ agent/squad sends out its overall situational awareness to all agents in its alliance. So this combat entity characteristic must be assumed to be at max in modeling scenarios.

MANA 5 (2006/2007) is a fully developed version of MANA which is now in use. Instead of the cell-based movement algorithms of previous MANA versions, a vector-based movement algorithm was used. A weighted vector is calculated towards targets of interest and a movement 'force' is applied to the agents. It was anticipated that a vector-based movement system will open up the flexibility to develop more intelligent agent behavior. For example, there is an interest in intelligent path finding for agents to navigate urban terrains without becoming lost or stuck in corners (McIntosh, Galligan, Anderson, & Lauren, 2008).

Expert Systems

An **expert system** is a computer application that performs a task that would otherwise be performed by a human expert. For example, there are expert systems that can diagnose human illnesses, make financial forecasts, and schedule routes for delivery vehicles. Some expert systems are designed to take the place of human experts, while others are designed to aid them. An expert system consists of these three components:

- Structure
- Knowledge base
  - Contains all the rules (rule-base) and most of the facts.
  - Rules are in the form of:
    - antecedent ⇒ consequent or if ⇒ then
- Inference engine

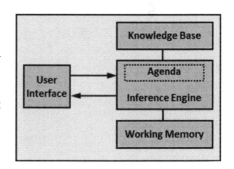

Figure 11-20. The structure of an expert system appears in the figure above.

The knowledge base contains all the rules and most of the facts about the system.

The inference engine controls overall execution of the rules. It searches through the knowledge base, attempting to pattern match facts or knowledge present in memory to the antecedents of rules. If a rule's antecedent is satisfied, the rule is ready to fire and is placed in the agenda. When a rule is ready to fire it means that since the antecedent is satisfied, the consequent can be executed. Salience is a mechanism used by some expert systems to add a procedural aspect to rule inferencing. Certain rules may be given a higher salience than others may, which means that when the inference engine is searching for rules to fire, it places those with a higher salience at the top of the agenda. Inferencing is to computers what reasoning is to humans.

CLIPS & Derivatives

CLIPS is a productive development and delivery expert system tool which provides a complete environment for the construction of rule and/or object based expert systems. The origins of the C Language Integrated Production System (CLIPS) date back to 1984 at NASA's Johnson Space Center. At this time, the Artificial Intelligence Section had developed over a dozen prototype expert systems applications using state-of-the-art hardware and software. CLIPS 6.2 was released in Spring 2002. CLIPS is now maintained independently from NASA as public domain software.

CLIPS is probably the most widely used expert system tool because it is fast, efficient and public domain. CLIPS incorporates a complete object-oriented language COOL for writing expert systems. CLIPS is written in C, extensions can be written in C, and CLIPS can be called from C. Its user interface more closely resembles that of the programming language LISP.

Gensym's Corporation's G2 Expert Systems Development Environment is a object-oriented development and deployment environment for managing and optimizing dynamic, complex decision support and control applications like UAV (ILS, 2009). G2 automates decisions in the demanding real-time applications ranging from monitoring NASA's spacecraft operations to detecting and solving problems in Ericsson's global telecommunications networks to coordinating Toyota's automobile production in its worldwide facilities. The U.S. Department of Energy uses G2-based application for planning the removal of hazardous waste from underground tanks located at DOE's Hanford Site next to the Columbia River in southeastern Washington (Capitol Reports, 2006).

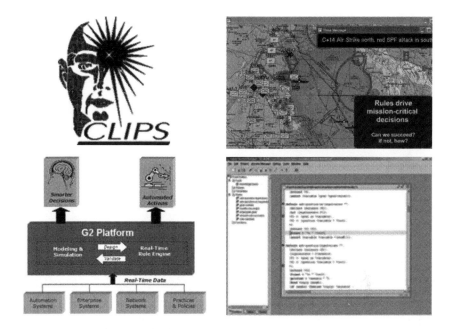

Figure 11-21. CLIPS and derivatives including G2 and JMACE

In use since 2001, Joint Military Art of Command Environment (JMACE) was built on Gensym's G2 environment.it is one of the simulation tools that the U.S. National Simulation Center (NSC), uses to provide accurate simulations of mission-critical operations. It serves as both a training tool, used to define new military training and exercise simulations, as well as a wargaming tool, used to simulate battle plans for upcoming offensives. It was used by the NSC in 2001 and 2002 to support US Army Command and General Staff College, US Central Command (Operation Iraqi Freedom) (Pawlowski, 2005).

In 2003, the United States Central Command (CENTCOM) utilized JMACE in important wargaming exercises. JMACE provided rule-driven simulations of different battle scenarios, taking into account the pre-defined rules of engagement. The application considered logistics, troop and equipment positions, and communications bandwidth, to name a few. JMACE was even able to factor in the element of surprise, using line of vision calculations and terrain mapping, and enemy resistance, based on years of experience from CENTCOM commanders. JMACE simulated the battle plan, allowing military commanders to test different situations and scenarios on a realistic, but bloodless battlefield.

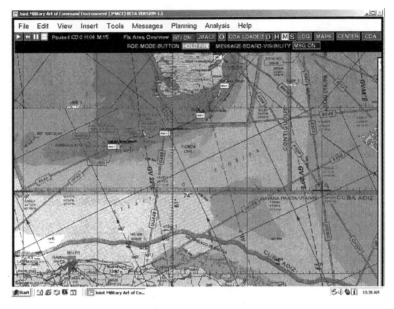

Figure 11-22. JMACE Sample Screen Shot

Military Entertainment Complex

Throughout the 1980s and in the early 1990s game developers and military personnel generally worked in isolation. Both groups were developing Wargames, but few in the gaming community were granted access to the resources (material, financial and human) of the armed forces.

In the late 90s, however, that trend changed dramatically. In 1997, the Marine Corps signed a deal with MÄK Technologies to create the first combat-simulation video game "to be co-funded and co-developed" by the Department of Defense and the entertainment industry. A year later, the Army signed a contract with MÄK to develop a sequel to its commercial tank simulation game "*Spearhead*" for use by the U.S. Army Armor Center and School and the Army's Mounted Maneuver Battle Lab (Turse, 2003). Since then, the entertainment industry and the armed forces have worked together to develop the following computer games and video games:

- "Tom Clancy's Rainbow Six: Rogue Spear" (2001)
- "Americas Army" (2002)
- "Rainbow Six: Raven Shield" (2003)

- "SOCOM II: U.S. Navy Seals" (2003)

America's Army (early versions)

America's Army (also known as AA or Army Game Project) is a tactical multiplayer first-person shooter owned by the United States Government and released as a global public relations initiative to help with U.S. Army recruitment.

The PC version, subtitled Recon, was first released on July 4, 2002. Subsequently Operations was first released on July 12, 2002. On November 6, 2003, version 2.0 of America's Army was released, with the full title of America's Army: Special Forces. The most current version 2 release, Coalition, debuted Dec 21, 2006 and has had many upgrades since Recon. It is financed through U.S. tax dollars and distributed for free. It was originally developed by the MOVES Institute at the Naval Postgraduate School and continues to use the Unreal Engine (Strickland, Networked Virtual Simulation, 2006).

Figure 11-23. America's Army 2.0 Startup Screen

In December 2003, a The Boston Globe columnist said "... *America's Army isn't just a time-wasting shoot-'em-up. It's full of accurate information about military training and tactics, intended to prepare a new generation of potential recruits. Amidst all the shouting drill sergeants and whistling bullets, some real education is going on. America's Army is a 'serious game,' part of a new wave of computer simulations that provide entertaining lessons about real world activities* (Bray, 2003)."

The Unreal Engine is a widely-used game engine developed by Epic Games. First illustrated in the 1998 first-person shooter game Unreal, it has been the basis of many games since, including *Unreal Tournament*, Tom Clancy's *Rainbow Six* 3: Raven Shield, Red Steel , and Gears of War. Although primarily developed for first-person shooters, it has been successfully utilized in a variety of genres, including 3rd-person stealth (Tom Clancy's Splinter Cell) and MMORPG (Vanguard: Saga of Heroes).

Its core written in C++, the Unreal Engine features a high degree of portability, supporting a plethora of platforms including the IBM PC compatibles (Microsoft Windows, GNU/Linux), Apple Macintosh (Mac OS, Mac OS X) and many consoles (Dreamcast, Xbox, Xbox 360, Playstation 2, Playstation 3). A great deal of the gameplay code is written in UnrealScript, a proprietary scripting language, and as such large parts of the gameplay can be modified without delving deep into the engine internals. Additionally, as with other middleware packages, the Unreal Engine also provides various tools to assist with content creation, both for designers and artists (Strickland, Networked Virtual Simulation, 2006).

Unreal Engine 2 (America's Army v1.0 ~ v2.x)

The sophomore version of the Unreal Engine got off to a rocky start with the mixed reviews for Unreal Tournament 2003. This generation saw the core code and rendering engine completely re-written and the new Unreal Ed 3 integrated. It also integrated the Karma physics SDK, which powered the vehicles in Unreal Tournament 2004. Many other engine elements were also updated, with improved and added support for the PlayStation 2 and the Xbox, respectively (Strickland, Networked Virtual Simulation, 2006).

America's Army 3

It was announced in early 2008 that the next version, America's Army 3, would be released in "fall 2008". Due to technical issues and problems with liaisons between the various departments the game, release was delayed and rescheduled for "some time in 2009". America's Army 3 entered beta testing in late 2008 and was released on June 17, 2009. It uses the Unreal 3 Engine and introduces other changes to how a player manages his account. Despite America's Army 3 being an entirely new game engine, there are still a number of similarities between America's Army and America's Army 3 which include the two remade and similar training. This version is said to put emphasis on graphical performance and on graphical flexibility to cover a greater range of PCs, as well as decreased size for the full version download (Recent News From The Development Team, 2007) (An Update From The Development Team, 2007). The game also features fictional weapons for the enemy as opposed to the Soviet and Warsaw Pact based weapons used in the previous versions. One day after the launch the civilian developers team contracted until game release were dismissed. Responsibility for future development of the game was passed on to another office at Redstone Arsenal.

Government applications

In 2005, the America's Army developers partnered with the Software Engineering Directorate and the Army's Aviation and Missile Research Development Engineering Center in Huntsville, Alabama, to manage the commercial game development process and use the America's Army platform to create government training and simulations. "America's Army has pushed to reuse the same elements for many purposes," said Colonel Wardynski, originator of the Game, "*We can build one soldier avatar and use it again and again. When we build something in America's Army, the U.S. government owns it completely ... and* [it] *can therefore be used for any application or use of the game. So costs keep going down*" (Testa, 2008). After AA went live, requests started coming in to use the game for purposes other than recruiting, such as training.

The partnership with SED, an Army software lifecycle management center, allowed the development team to repurpose the commercial software to meet the needs of Soldiers preparing for deployment. SED engineers developed customized applications used by many different Army and government organizations including the JFK Special Forces School and the Army's Chemical School. They are used to provide training in use of rare equipment such as PackBot robots, CROWS, and Nuclear Biological Chemical Reconnaissance Vehicles.

Full Spectrum Warrior

Created by Pandemic, Sony Imageworks and the Institute for Creative Technologies at USC (ICT) for STRICOM, *Full Spectrum Warrior* is a squad-based training game that runs on the Microsoft X-Box. It places the student in the role of a light infantry squad leader. A version of the game has also been released commercially. The game has been adapted by psychologists to assist veterans from Iraq overcome the effects of post-traumatic stress disorder (Virtual Iraq, 2009).

Full Spectrum Command is a company command level game, created by the Institute for Creative Technologies at USC (ICT) and Quicksilver,. As the commander of a U.S. Army light infantry Company, the student must interpret the assigned mission, organize his force, plan strategically, and coordinate the actions of about 120 soldiers under his command. The game is used in week 8 of the Infantry Captains' Career Course on battlefield synchronization at Fort Benning. Players create tactics and command their company from multiple perspectives.

Figure 11-24. *Full Spectrum Warrior* screenshot

Development

In 2000, the US Army Science & Technology community was curious to learn if commercial gaming platforms could be leveraged for training. Recognizing that a high percentage of incoming recruits had grown up using entertainment software products, there was interest in determining whether software game techniques and technology could complement and enhance established training methods.

Having established a US Army University Affiliated Research Center (the Institute for Creative Technologies) in 1999 for the purpose of advancing virtual simulation technology, work began in May 2000 on a project entitled C4 under ICT Creative Director James Korris with industry partners Sony Imageworks and their team-mate, Pandemic Studios, represented by co-founders Josh Resnick and Andrew Goldman.

At the time, there was a great deal of interest in leveraging the stability, low cost and computational/rendering power of the new generation of game consoles, chiefly Sony's PlayStation 2 and Microsoft's Xbox, for training applications. Legal restrictions on the PlayStation (using the platform for a military purpose) combined with the default Xbox configuration "persistence" (i.e. missions recorded on the embedded hard drive for after-action review) led to the final selection of the Xbox platform for development (Perry, 2004).

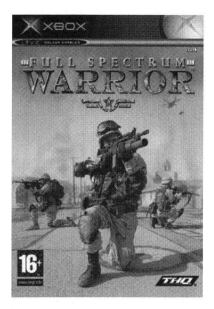

A commercial release of the game was required for Xbox platform access. The team, however, quickly concluded that a viable entertainment title might differ from a valid training tool. The exaggerated physics of entertainment software titles, it was believed, could produce a negative training effect in the Soldier audience. Accordingly, the team developed two versions of the game. The Army version was accessible through a static unlock code; the entertainment version played normally.

Andrew Paquette, a former Sony art director, says the companies were so focused on creating a best-selling game that they cut corners on the Army version. As a result, the urban scenes are not as accurate as they should be. The designers *"pretty much disregarded the Army's concerns,"* said Paquette. *They wanted to make money on the commercial version"* (Adair, 2005). Lt. Col. Jim Riley, chief of tactics at the Army's infantry school at Fort Benning, Ga., says his school rarely uses the game because it does not offer a realistic simulation of urban combat. *"It's not accurate enough,"* he said (Adair, 2005).

The most radical decision in the game's development was to limit first-person actions to issuing orders and directions to virtual Fire Teams and Squad members (see Gameplay). Given the popularity of the first-person shooter genre, it was assumed that all tactical-level military gameplay necessarily involved individual combat action. The application defied conventional wisdom, winning both awards and commercial acceptance. The game's working title evolved to C-Force (2001) and ultimately Full Spectrum Warrior (2003).

As work progressed on Full Spectrum Warrior, ICT developed another real-time tactical decision-making game with Quicksilver Software entitled Full Spectrum Command for the US Army's Infantry Captains Career Course, with the first-person perspective of a Company Commander. As the application was designed to play on a desktop PC (unlike the Xbox), no commercial release was necessary. Full Spectrum Command gave rise to a sequel developed for the US Army and Singapore Armed Forces (version 1.5). A related ICT/Quicksilver title, Full Spectrum Leader, simulates the first person perspective of a Platoon Leader.

Full Spectrum Warrior relates to the Army's program of training soldiers to be flexible and adaptable to a broad range of operational scenarios.

SOAR Architecture

The State Operator and Response (SOAR) is a symbolic cognitive architecture, created by John Laird, Allen Newell, and Paul Rosenbloom at Carnegie Mellon University, now maintained by John Laird's research group at the University of Michigan. It is both a view of what cognition is and an implementation of that view through a computer programming architecture for Artificial Intelligence (AI). Since its beginnings in 1983 and its presentation in a paper in 1987 (Rosenbloom, Laird, & Newell, 1993), it has been widely used by AI researchers to model different aspects of human behavior (Lehman, Laird, & Rosenbloom, 2006).

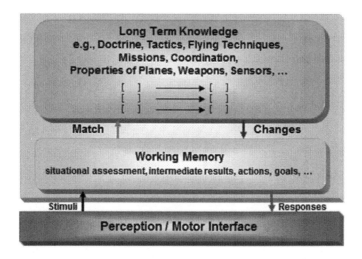

Figure 11-25. SOAR Architecture diagram

Specifically, an agent in the world owns goals which are applied to a problem space. This application changes that agent's current state through the application of operations to the world. For example, an aircraft may have the goal of shooting down an enemy aircraft. It must do this within the problem space of what it understands about air-to-air combat. The aircraft makes decisions about which actions to take based on the problem space, its current state, and the operators available to it. The selection and application of a specific operator, like a Pop-and-Roll maneuver, will change the aircraft's state values. This places it in a different condition for its next decision. A series of these decisions should lead to the accomplishment of the stated goal.

In order to perform these reasoning processes, the SOAR system implements Working Memory about its current situation. Moreover, it implements Long-Term memory about past experiences or its standing knowledge base and a Perception/Motor Interface which converts its perceptions of the real world into actions of the agent. It implements a Decision Cycle which evaluates all the options available and selects those that are feasible. It implements Impasses, which capture situations in which more knowledge is necessary for making a decision; and finally, Chunking, which creates new associations in the knowledge base, allowing additional decision making.

SOAR has been used extensively for military problems. Most notably, the agents of Soar-intelligent forces (Soar-IFOR) have been participants in simulated theaters of war (STOW-E in 1995, STOW-97 in October 1997). These agents have simulated the behavior of fixed-wing and rotary-wing pilots on combat and reconnaissance missions, competing favorably with human pilots.

TacAir-Soar

Soar Technology's *TacAir-Soar* offers fully autonomous synthetic forces with high-fidelity, realistic behaviors for a wide range of aircraft and missions. Users of TacAir-Soar realize direct cost savings due to reduced manpower requirements for training exercises or experiments.

TacAirSoar entities are not limited to performing their pre-briefed missions or following a script. They coordinate their actions through shared goals, planning and explicit communication. Even though each entity is autonomous, it does not act in isolation. Individual entities coordinate their actions using existing doctrine and C4I systems. As the mission develops, entities may change roles dynamically.

The key to simulation is believability, and the behaviors of TacAirSoar entities are more realistic than any currently available system not controlled by a human. Because TacAirSoar is based on a model of human cognition, it not only creates reasonable behaviors, but also chooses actions and makes decisions for similar reasons that a human would. TacAirSoar's knowledge base consists of over 7500 rules, making it among the largest fielded expert systems (Laird, Jones, & Nielsen, 1998).

Tank Soar

In *Tank Sore*, I created a set of rules for the Green Tank to use as it moves around on the battlefield. These rules include when to go forward or backward, when to turn (it cannot run over the trees), when to turn its radar on, when to turn its shields off, when to fire missiles, when to search for ammunition, when to search for energy, etc. The red, blue, and yellow tanks do nothing but provide target for the Green Tank's missile, and serves as hostile things in the environment that the Green Tank should respond to (i.e., turn on/off shields, turn on/off radar, avoid, engage, etc.).

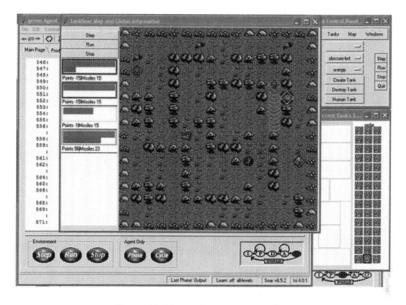

Figure 11-26. Tank SOAR screenshot

Artificial Problem Solvers: Swarm Intelligence

"Any attempt to design algorithms or distributed problem-solving devices inspired by the collective behavior of social insect colonies and other animal societies."

- (Bonabeau, Dorigo, & Theraulaz, 1999)

Swarming Characteristics

The concept of stigmergy is used to describe the indirect communication taking place among individuals in social insect societies. Stigmergy was first observed in social insects. For example, ants exchange information by laying down *pheromones* (the trace) on their way back to the nest when they have found food. In that way, they collectively develop a complex network of trails, connecting the nest in the most efficient way to the different food sources. When ants come out of the nest searching for food, they are stimulated by the pheromone to follow the trail towards the food source. The network of trails functions as a shared external memory for the ant colony. In computer science, this general method has been applied in a variety of techniques called ant colony optimization, which search for solutions to complex problems by depositing "virtual pheromones" along paths that appear promising.

Guy Theraulaz and Eric Bonabeau described stigmergy: Stigmergy is a class of mechanisms that mediate animal-animal interactions. Its introduction in 1959 by Pierre-Paul Grasse made it possible to explain what had been until then considered paradoxical observations. In an insect society individuals work as if they were alone while their collective activities appear to be coordinated. Grasse stated that

"The coordination of tasks and the regulation of constructions does not depend directly on the workers, but on the constructions themselves. The worker does not direct his work, but is guided by it. It is to this special form of stimulation that we give the name STIGMERGY (stigma, goad; ergon, work, product of labor = stimulating product of labor)."

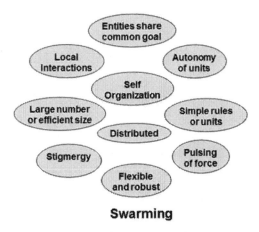

Swarming

Figure 11-27. Characteristics of swarms

Nature has given us several examples of how minuscule organisms, if they all follow the same basic rule, can create a form of collective intelligence on the macroscopic level. As we have seen, colonies of social insects perfectly illustrate this model which greatly differs from human societies. This model is based on the co-operation of independent units with simple and unpredictable behavior (Waldner, Inventer l'Ordinateur du XXIème Siècle, 2007). They move through their surrounding area to carry out certain tasks and only possess a very limited amount of information to do so. A colony of ants, for example, represents numerous qualities that can also be applied to a network of ambient objects. Colonies of ants have a very high capacity to adapt themselves to changes in the environment as well as an enormous strength in dealing with situations where one individual fails to carry out a given task. This kind of flexibility would also be very useful for mobile networks of objects which are perpetually developing. Parcels of information that move from a computer to a digital object behave in the same way as ants would do. They move through the network and pass from one knot to the next with the objective of arriving at their final destination as quickly as possible (Waldner, Nanocomputers and Swarm Intelligence, 2008).

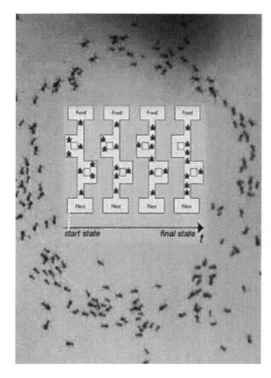

The main quality of the colonies of insects, ants or bees lies in the fact that they are part of a self-organized group in which the keyword is simplicity.

Every day, ants solve complex problems due to a sum of simple interactions, which are carried out by individuals.

The ant is, for example, able to use the quickest way from the anthill to its food simply by following the way marked with pheromones.

Figure 11-28. When a colony of ants is confronted with the choice of reaching their food via two different routes of which one is much shorter than the other, their choice is entirely random. However, those who use the shorter route move faster and therefore go back and forth more often between the anthill and the food

Routing in Networks

In computer science, Artificial Ants stand for multi-agent methods inspired by the behavior of real ants. The pheromone-based communication of biological ants is often the predominant paradigm used (Nicolas, Frédéric, & Patrick, 2010). Combinations of Artificial Ants and local search algorithms have become a method of choice for numerous optimization tasks involving some sort of graph, e. g., vehicle routing and internet routing. The burgeoning activity in this field has led to conferences dedicated solely to Artificial Ants, and to numerous commercial applications by specialized companies such as AntOptima. As an example, Ant colony optimization (Dorigo & Gambardella, 1997) is a class of optimization algorithms modeled on the actions of an ant colony. Artificial 'ants' (e.g. simulation agents) locate optimal solutions by moving through a parameter space representing all possible solutions. Real ants lay down pheromones directing each other to resources while exploring their environment. The simulated 'ants' similarly record their positions and the quality of their solutions, so that in later simulation iterations more ants locate better solutions (Dorigo & Stützle, 2004). One variation on this approach is the bees algorithm, which is more analogous to the foraging patterns of the honey bee, another social insect.

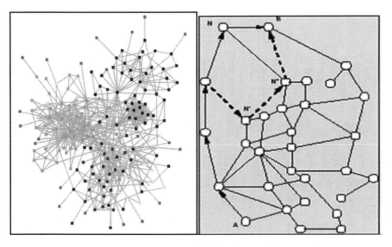

Figure 11-29. A complex network (left) and a simpler network (right)

Sugarscape

Joshua Epstein and Robert Axtell begin the development of a "bottom up" social science in a computer model. Their program, named *Sugarscape*, simulates the behavior of artificial people (agents) located on a landscape of a generalized resource (sugar). Agents are born onto the Sugarscape with a vision, a metabolism, a speed, and other genetic attributes. Their movement is governed by a simple local rule: "look around as far as you can; find the spot with the most sugar; go there and eat the sugar." Every time an agent moves, it burns sugar at an amount equal to its metabolic rate. Agents die when they burn up all their sugar. A remarkable range of social phenomena emerges. For example, when seasons are introduced, migration and hibernation can be observed. Agents are accumulating sugar at all times, so there is always a distribution of wealth.

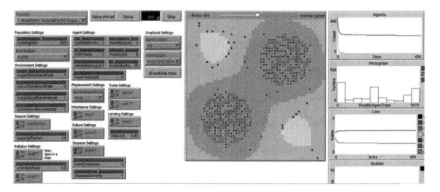

Figure 11-30. A version of *SugarScape*, as presented in "Growing Artificial Societies" by Epstein and Axtell. It builds on Owen Densmore's NetLogo community model to encompass all rules discussed in GAS with the exception of the combat rule (although trivial to include, it adds little value to the model).

Next, Epstein and Axtell attempted to grow a "proto-history" of civilization. It starts with agents scattered about a twin-peaked landscape; over time, there is self-organization into spatially segregated and culturally distinct "tribes" centered on the peaks of the *Sugarscape* (Epstein & Axtell, 1996). Population growth forces each tribe to disperse into the sugar lowlands between the mountains. There, the two tribes interact, engaging in combat and competing for cultural dominance, to produce complex social histories with violent expansionist phases, peaceful periods, and so on. The proto-history combines a number of ingredients, each of which generates insights of its own. One of these ingredients is sexual reproduction. In some runs, the population becomes thin, birth rates fall, and the population can crash. Alternatively, the agents may over-populate their environment, driving it into ecological collapse.

When Epstein and Axtell introduce a second resource (spice) to the Sugarscape and allow the agents to trade, an economic market emerges. The introduction of pollution resulting from resource-mining permits the study of economic markets in the presence of environmental factors.

EINStein

ISAAC (Irreducible Semi-Autonomous Adaptive Combat) was developed as a simple "proof-of-concept" model to illustrate how combat can be viewed

227

as an emergent self-organized dynamical process. It introduced the key idea of building combat "up from the ground up" by using complex adaptive agents as primitive combatants and focusing on the global co-evolutionary patterns of behaviors that emerge from the collective nonlinear local dynamics of these primitive agents (Ilachinski, An Artificial-Life Approach to War, 2001).

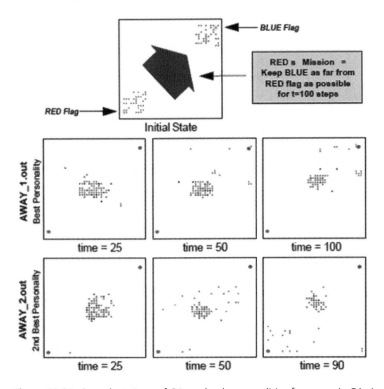

Figure 11-31. Snapshot views of GA-evolved personalities for scenario GA_1

EINSTein builds upon and extends ISAAC into a bona-fide research tool for exploring self-organized emergent behavior in combat (Ilachinski, Towards a Science of Experimental Complexity, 2000). Some of the features planned for EINSTein include:

- A fully integrated Windows 95 GUI front-end
- An object-oriented C++ code base (compared to ISAAC's plain vanilla ANSI-C source)
- Integrated natural terrain maps and terrain-based adaptive decision-dynamics

- Context-dependent and user-defined agent behaviors (i.e. personality scripts)

- On-line genetic algorithm, neural-net, reinforcement-learning, and pattern recognition toolkits

- On-line data collection and multi-dimensional visualization tools

- On-line chaos-data/time-series analysis tools

- On-line mission-fitness co-evolutionary landscape profilers

Developed by Andy Ilachinski, Tactical Analysis Team, Operations Evaluations Group

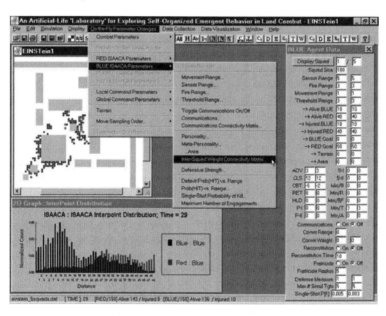

Figure 11-32. Screenshot of EinSTEIN

Interactive "Tool Box" / Laboratory

EINSTein is the centerpiece of an ambitious on-going project designed to weave together several inter-related methodological strands (see figure). First and foremost, EINSTein will serve as an interactive "tool box" (or conceptual laboratory) for the general exploration of combat as a complex adaptive system. As such, analysts will have an opportunity to play multiple "What If?" scenarios and to experiment with fundamental issues of the dynamics of war.

Secondly, EINSTein will provide a collection of on-line data-collection/data-analysis and pattern recognition tools; this will mark the first time that many of the tools commonly found in the study of nonlinear dynamical, and complex adaptive, systems will be available in the context of military combat analysis.

Thirdly, EINSTein will have a data visualization package to facilitate the exploration of multiple high-dimensional co-evolutionary fitness landscapes, and to foster the development of an intuition of the overall combat phase space. With embedded genetic algorithm, neural-net, and reinforcement learning based adaptive learning toolkits, EINSTein will be an ideal testbed for developing future tactical decision aids.

Finally, the long-term vision is to extend EINSTein to become a remote dynamic real-time "combat engine" that assimilates, processes, and automatically analyzes the evolutionary outcomes and consequences of user-defined scripts, tactics and doctrine. In short, a fully realized intelligent-agent-based laboratory for exploring collective self-organized emergent behavior in combat.

MARSS

NPS developed the Multi-Agent Robot Swarm Simulation (MARSS) for modeling the behavior of swarms of military robots. MARSS contains state, sensing and behavioral model building tools that allow a wide range of complex entities and interactions to be represented. It is a model-building tool that draws theory and ideas from agent-based simulation, discrete event simulation, traditional operations research, search theory, swarm theory, and experimental design. MARSS is designed to explore the effect individual behavioral factors have on swarm performance. The performance response surface can be explored using designed experiments, or using genetic evolution. Using MARSS, a model was developed to investigate the effects of increasing behavioral complexity for a search scenario involving a swarm of 25 Micro Air Vehicles searching for four mobile tanks in a predefined region. Preliminary results show agreement between theoretical and simulated search scenarios for simple searchers (Dickie, 2002).

MARSS features

The features of MARSS include:

- developed using Java technologies,
- adaptable to a wide range of models,
- linked 2D and 3D views of the model,
- rapid scenario development using XML,
- multi-run capability,
- controlled random seed use,
- physical based modeling of movement,
- uses Simkit (event based simulation model),
- automated designed experiments, and
- easily deployable.

Sample Projects

Created by the MOVES institute at the Naval Postgraduate School, America's Army is an Army awareness and recruiting tool. It is a squad-based first-person shooter game consisting of "basic training," plus a series of team-based "missions" which involve operations, Special Forces, and Combat Medic specialties.

Created by Will Interactive, Anti-Terrorism Force Protection is an interactive training program to train commanders to make decisions related to their command's anti-terrorism posture.

Created by Forterra, Inc. for RDE COM, Asymmetrical Warfare Virtual Training Technology is an R&D project about building a new type training approach on Forterra Inc.'s Virtual World platform.

Created by MÄK Technologies, *BC-2010* is designed to support Army battalion and brigade commands and their staff officers in preparing operation orders, the game is a military tactical trainer that allows commanders and their staff officers to practice their planning and execution skills within a simulated environment. Being used at Ft. Leavenworth, Ft. Knox, the USMA, the 35th infantry division National Guard, and has been field tested in the 1st Infantry Division in Europe.

A 12-hour basic training simulation for navy recruits, it is further developing and replacing a very basic "jury-rigged" simulation developed by petty officers. In battle Station 21, 352 Navy recruits at a time will "man" a ship and draw on all they've learned to handle a "very, very real and visceral environment."

A staff and soldier training program that can be modified to meet each soldier's needs. Will expedite the training process of new recruits through simulated exercises "instead of [having to take] a bunch of guys out into the woods." From the University of Texas Institute for Advanced Technology. Part of the University XXI program, a joint project between the University, Texas A&M University and the U.S. Army, which began in 1999.

Created by Breakaway Games based on technology created by Booz-Allen, *Entropy-Based Warfare* is a game that has been used by the OSD and the Army War College.

Created by Cornerstone Industry for the Army National Guard, Guard Force is a real-time strategy game focused on the Guard's combat and non-combat missions, from counterinsurgency to rescue.

Chapter 12. Multi-Resolution Modeling

This lecture describes the use of multiple levels of model resolution in order to address questions at different levels of detail and as a mechanism for tying simulations together.

Multi-Resolution Modeling

Multi-Resolution modeling is the practice of representing the synthetic world with models of varying levels of resolution. Often, models that represent different levels of resolutions are not in harmony. The challenge is creating a distributed simulation environment that simultaneously and harmoniously supports models operating at different levels of resolution. This can be accomplished by integrating two different simulations using the interoperability techniques described earlier, like DIS or HLA. Alternatively, is can be can be done by creating multiple modules for the same model that can be exchanged depending upon the needs of the study or training.

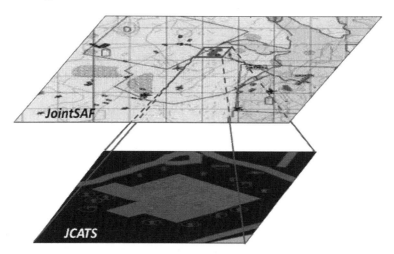

Figure 12-1. JCATS is an entity-level simulation, while JointSAF aggregates the entities to battalion or above units.

Another term that is developing is Multi-Resolution Multi-Perspective Modeling (MRMPM). This adds to MRM in that the data must be modeled in multi-resolution, but it must also be presented from multiple perspectives. The users of the system must be able to view and analyze model output at any of the resolution levels that are presented.

Common Resolution Levels

Half of a century of model development has led to an environment in which models usually fall into four different categories.

The first category is the **Theater** or **Campaign** model in which aggregate units are controlled via scripted missions. These models encompass all aspect of a conflict that may span the entire globe. This type of model is often used for force-structure studies.

The second category is the **Mission** or **Battle** model which focuses on specific packages of forces, usually limited to a specific geographic region. It may represent assets as individual vehicles or as aggregate units, and will often combine the two. This type of model is used in both the analytical and the training communities.

The **Engagement** model is used to evaluate individual weapon systems or to train crews to use them. It limits the scenario to select few units and focuses on a relatively small geographic area.

Engineering models study a single weapon system or the sub-systems within in it. These models may be used to evaluate design alternatives for new sensors, wing structures, or warhead configuration.

Aggregation

Aggregation is the ability to group entities while preserving the effects of entity behavior and interaction while grouped. An aggregated model is one in which many detailed elements of a process are combined into and examined as a single entity. Thus, a model which treats a division as an entity in theater-level combat has aggregated platoons, scout patrols, fire-support batteries, companies, battalions, and brigades into the entity called a division and is therefore an aggregated model (Strickland, Mathematical Modeling of Warfare and Combat Phenomenon, 2011).

Military simulations are often differentiated based on their inherent level of abstraction. If the primary objects represented in the simulation are collections of doctrinally identifiable military assets, e.g. a tank battalion, then the simulation is referred to as an aggregate-level simulation. If the primary objects represented by the simulation are singular military objects, e.g. a tank, a soldier, the simulation is referred to as an entity-level simulation.

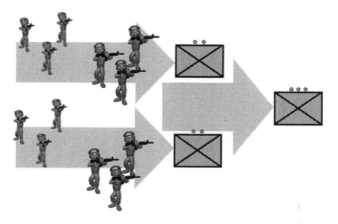

Figure 12-2. The aggregation process: going from individual entities, to unit entities (squad) and then to an higher unit entity (platoon)

Effects of Aggregation

The US military trains soldiers at many different levels—gunnery, team, company, battalion, corps, and echelons above corps to name a few. Each of these is supported by a different set of tools, some of which are simulations. Historically, each area has developed its own tools, but recently the trend has been to use the same tools at multiple levels, or to join multiple levels together and expect them to interoperate consistently. These connections are hindered by the capabilities of the tools and simulations in the field, the fundamental differences in the problem spaces, and scaling issues of the computer systems and operational techniques. This has resulted in the need to design simulations and interfaces that are more scaleable than those of the past.

Proper Aggregation

Proper Aggregation is a somewhat ad hoc process by which the modeler decides that the members of some collection of observable things may be considered identical. Proper Aggregation generally has associated with it a Resolution of some sort (Fowler, 1996).

The test of a Proper Aggregation is largely statistical. For example, there are two components to Proper Aggregation for an attrition thing. How the thing attrits other things; and how it is attrited.

In general, we shall refer to the things that have been Properly Aggregated together as elements. This terminology is intended to reinforce the essentially fundamental nature of these class-aggregated things within the modeling context. In Classical Lanchester Theory, Lanchester uses the term "units", but we have chosen not to use this terminology to avoid confusion with the military organizational use of this term.

Example: Suppose a group of tanks consists of equal numbers of M1A1's and M1A2's.

Tank	Primary Armament
M1A1	105 mm Main Gun
M1A2	120 mm Main Gun

The guns have different lethality properties. If the difference as small in relation to the degree of resolution sought in the model, they may be "properly aggregated."

Formal Aggregation

Proper Aggregation is an aggregation of definition while Formal Aggregation is an aggregation of transformation.

Formal aggregation takes as a fundamental that the aggregation process must not only conserve some observables of the things being aggregated, but it must conserve the mathematical symmetry properties of those observables. That is, if we aggregate from a heterogeneous Lanchester picture to a homogeneous Lanchester picture, the mathematical process of that aggregation must conserve the symmetry properties of homogeneous Lanchester (Strickland, Mathematical Modeling of Warfare and Combat Phenomenon, 2011).

$$\frac{dM_i}{dt} = \sum_{j=1}^{N_R} \alpha_{ij}(\ )N_j \text{ and } \frac{dN_j}{dt} = \sum_{j=1}^{M_b} \beta_{ji}(\ )m_i \rightarrow \frac{d\vec{F}}{dt} = \vec{\Gamma} \cdot \vec{F}$$

Extensive Aggregation

Extensive aggregation occurs when the heterogeneous forces being aggregated are fighting uncoupled engagements (Strickland, Mathematical Modeling of Warfare and Combat Phenomenon, 2011). This can be represented in matrix form as

$$\vec{\Gamma} = \begin{bmatrix} 0 & 0 & -x & 0 \\ 0 & 0 & 0 & -x \\ -x & 0 & 0 & 0 \\ 0 & -x & 0 & 0 \end{bmatrix}$$

Intensive Aggregation

Intensive aggregation occurs when the heterogeneous forces being aggregated are fighting a coupled engagement (Strickland, Mathematical Modeling of Warfare and Combat Phenomenon, 2011). This can be represented in matrix form as

$$\vec{\Gamma} = \begin{bmatrix} 0 & 0 & -x & -x \\ 0 & 0 & 0 & -x \\ -x & -x & 0 & 0 \\ 0 & -x & 0 & 0 \end{bmatrix}$$

The governing attrition differential equations are

$$\frac{d\vec{F}}{dt} = \vec{\Gamma} \cdot \vec{F}$$

Which has the explicit eigen-solution $\vec{F}(t) = \bar{e} \cdot \vec{T}(t) \cdot e^{-1}\vec{F}(0)$

Where

$$\vec{T}(t) = \begin{bmatrix} e^{\lambda_1 t} & 0 & 0 \\ 0 & e^{\lambda_2 t} & 0 \\ 0 & 0 & e^{\lambda_3 t} \end{bmatrix}$$

The eigen-solution methodology preserves the mathematical symmetry that is required. In simple terms, the mathematical form of the end result of aggregation has the same mathematical properties as homogeneous Lanchester (Strickland, Mathematical Modeling of Warfare and Combat Phenomenon, 2011).

Hierarchical Attrition Algorithms

Most combat models have the characteristic that higher-resolution models feed them with data.

- **High-resolution models** need PH's and PK|H's which are usually supplied by AMSAA and ARL engineering models.

- Medium and low-resolution models usually depend on kill rates from high-resolution models or occasionally on engineering models.

- Some low-resolution models depend on kill rates generated by medium-resolution models, e.g., ATCAL (Attrition Calibration) links the medium-resolution COSAGE with the low-resolution CEM.

Figure 12-3. Most combat models have the characteristic that higher-resolution models feed them with data

A hierarchy of ground combat models was proposed and partially built by the Army in the 1970s and 1980s, but never reached completion because of enormous technical difficulties

- The vision was a semi-automated way of getting input data for low-res models from high-res models, so low-res modelers would not have to request special runs from high-res modelers (usually in different organizations).

- Too much variation in scenarios, force structures, weapons, and postures meant that libraries of high-res results were too hard to build.

- Additionally, organizations resisted getting "answers" from "black-box" models they could not influence.

ATCAL is one of the few examples of an existing formal link between models of differing resolution for the purpose of generating input data from the high-res model for the low-res model. Both models were built and are run by the same organization, contrary to what was envisioned.

Hierarchical Attrition Algorithms - ATCAL

ATCAL is an aggregated attrition model developed in the early 1980.s by the Center for Army Analysis for use in the CEM and FORCEM theater level simulations. ATCAL consists of a number of equations that can be used to compute attrition (if values for several input parameters are known). The same equations can be used, .backwards to determine values for the parameters from the output of a higher resolution division level model such as COSAGE. Thus ATCAL contains both an attrition model and a consistent calibration procedure.

The attrition equations of ATCAL are heterogeneous, computing casualties for firer-target pairings of weapon system types. There are two basic attrition equations in the methodology, one for point-fire weapons effects and one for "area fire."

The medium-resolution model generates three parameters (for point fire)

P_{ij}: Probability of kill of target j by firer i

A_{ij}: Availability of target j to firer i

$RATE_i$: shots per unit time by firer i

Care must be taken to ensure that all Blue firer types have an opportunity to fire at all Red target types and vice versa, so that all P_{ij} and A_{ij} have values.

The Resulting Aggregated Attrition Parameters

The following calibrated attrition parameters are carried over into the aggregated attrition model:

1. The target availability parameters, A_{ijk}, frontage independent.
2. The engagement ranges $RANGE_{ij}$, for scaling the target availability parameters.
3. The kills per round, P_{ijk}.
4. The average number of vehicles, AN_k
5. The firing capability parameters, $RATE_{ij}$.

High-Resolution Parameters:

A_{ijk} = the target availability is the fraction of the time that a single particular target of type k can be fired upon by a particular type i vehicle using weapon j. This availability, which can be considered to be averaged over all targets of type k and firers of type i, is a model input whose values are determined by the calibration procedure.

P_{ijk} = kills per round when a vehicle of type i uses its weapon of type j to engage a target of vehicle type k. This lethality measure is an input parameter to the attrition computation.

AN_k = the average number of type k vehicles alive during the battle. Since ATCAL does not compute in time steps, but rather assesses the entire engagement at once, we need the averages, AN_k, to determine the number of firing type k vehicles throughout the battle. ATCAL assumes an exponential decrease in weapon count during the battle.

$RATE_{ij}$ = the total number of shots that a single type j weapon on a type i vehicle is capable of firing during the entire duration of the battle (assuming targets are always available).

F_{ijk} = the number of shots fired by weapons of type j on type i vehicles at type k target vehicles. This is computed by the attrition model as the result of its fire allocation equation.

(DN_k) = the killer-victim scoreboard is the detailed result of the ATCAL attrition equation. It shows casualties to type k vehicles by the systems (i,j) that caused them. The casualties are computed as:

$$(DN_k)_{ij} = F_{ijk} * P_{ijk}$$

Development of the Point Fire Attrition Equation

Suppose that we have values for the target availability fractions, A_{ijk}, and the average number of targets, AN_k.

1. Assume that availability of each of the AN_k targets is independent of availability for the other type targets. Then the fraction of the time that no type k target is available for a particular firer of type i,j is

$$\left(1 - A_{ijk}\right)^{AN_k}$$

and thus the fraction of the time that a type k target is available is

$$1 - \left(1 - A_{ijk}\right)^{AN_k}$$

2. Assume that a firing weapon will shoot at the highest priority target available. Then, if type k is the highest priority for i,j, we can compute the number of shots fired at vehicles of type k by all weapons of type j on vehicles of type i as

$$F_{ijk} = AN_i \, RATE_{ij} \left[1 - \left(1 - A_{ijk}\right)^{AN_k}\right]$$

Now consider the case where type k is not necessarily the highest priority target.

3. Assume that availability of targets of type k is independent of availability of any other target type k'. Then, if firing at type k targets goes on only during the fraction of the time when no higher priority target is available, we can compute the general form for

$$F_{ijk} = AN_i \, RATE_{ij} \left[1 - \left(1 - A_{ijk}\right)^{AN_k}\right] \prod_{\text{all } k} \left(1 - A_{i'jk}\right)^{AN_{k'}}$$

where the subscript k' for the product ranges over all target types whose priority is higher than the priority of target type k.

Finally, the attrition to type k targets caused by all firers of type i, j is given by

$$(DN_k)_{ij} = P_{ijk} \cdot F_{ijk}$$
$$= P_{ijk} \, AN_i \, RATE_{ij} \left[1 - \left(1 - A_{ijk}\right)^{AN_k}\right] \prod_{\text{all } k} \left(1 - A_{i'jk}\right)^{AN_{k'}}$$

This is the basic point-fire attrition equation for the ATCAL attrition model. The attrition equation combines features of both the Lanchester Square Law and the Lanchester Linear Law (Caldwell, Hartman, Parry, Al Washburn, & Youngren, 2000).

What is Multi-Resolution Modeling?

Resolution is the level of detail at which system components and their behaviors are depicted. While fairly standard, this definition of resolution is ambiguous because each of the components of a model may have its own resolution. For example, high resolution may refer to including fine-grained entities such as individual tanks rather than tank companies. In addition, there is richer depiction of the entities (e.g., aspect-dependent vulnerability as well as position, velocity, and rate of fire), a richer depiction of how entities' characteristics depend on each other, a more detailed description of processes such as attrition and movement, or fine-grained scales in space or time. In comparing two models A and B, it is often the case that model A has lower resolution in some respects and higher resolution in other respects. Indeed, some "high-resolution models" have remarkably low-resolution depictions of important phenomena.

We define multi-resolution modeling (MRM) as

1. building a single model with alternative user modes involving different levels of resolution for the same phenomena;

2. building an integrated family of two or more mutually consistent models of the same phenomena at different levels of resolution; or both.

MRM really emerged from the practice of ***disaggregation*** used to connect aggregate and entity level simulations. In essence, disaggregation should "undo" aggregation. The core challenge for disaggregation is to convert a single lumpy object into a number of more finely represented objects. Disaggregation may be a bit more challenging than aggregation.

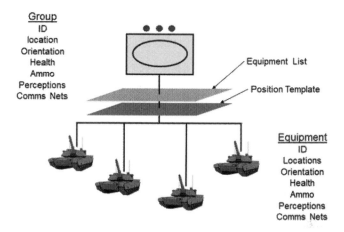

Figure 12-4. Process of Disaggregation (Strickland, Multiresolution Modeling, 2005)

This requires the identification of a list of entities contained in the aggregate. Some aggregate models include this list in the aggregate structure, while others do not. When not available, some of the attributes of the aggregate must be used as keys to look-up an equipment list is a separate database.

Once the equipment list is created, a template is usually applied to select a unique location for each piece of equipment. Most models use static templates that place equipment in positions based on the mission or status of the aggregate and the type of the entity equipment.

MRM: Vehicle Location and Health

The most common illustration of Multi-Resolution Modeling is joining an Engagement level model with a Mission level model. When this occurs, the primary data transformation is in the placement and health of individual vehicles in the engagement simulation such that they are consistent with the same units in the mission Simulation.

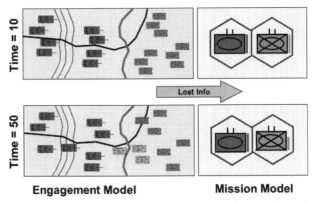

Figure 12-5. Vehicle Location and Health: which three IFV's are lost at the aggregate-level at Time = 50?

The MRM Problem

"It would prove quite fruitful for the Department of Defense to sponsor a number of studies ... In which workers go through real-world problems in considerable detail, attempting to design MRM models and recording their insights and methods"

The simulations are fundamentally different:

- JTLS is time stepped, JCATS is event stepped
- JTLS is hex-based, JCATS uses elevation posts
- JTLS adjudicates conflict using (mostly) Lanchestrian equations, JCATS uses $P_h P_k$
- JTLS uses aggregate units, JCATS uses individual weapons systems

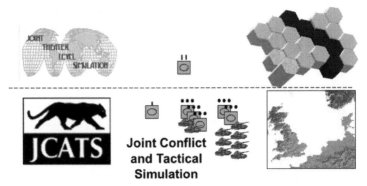

Figure 12-6. JCATS-JTLS, © 2002, the MITRE Corporation. All Rights Reserved (used with permission)

The Joint Conflicts and Tactical Simulation (JCATS) model and the Joint Theater Level Simulation (JTLS) model represent a good example of the problems posed by MRM. These models are fundamentally different in several respects including time and event stepping, terrain representation, attrition processes, and levels of aggregation.

Aggregate and Entity

MITRE has conducted experiments in joining aggregate and entity simulations. This one is in SIW paper 02F-SIW-034 (Prochnow, 2002).

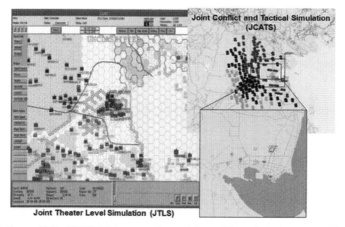

Figure 12-7. JCATS-JTLS screen shots. Adapted from (Prochnow, 2002)

In initial implementation, JCATS reflected everything in the JTLS playbox at entity level (build one). Since each of the battalions in JTLS corresponded to hundreds of entities in JCATS, JCATS was displaying tens of thousands of entities. Since the representations at a distance of more than about 50 km from the urban area were unnecessary for the scenario, a "box" was defined in JCATS outside of which no JTLS objects would be reflected.

Within this box, a second box (build 2) was added which formed the location for automatic transfer of object ownership for (some) air objects and the approximate location for manual hand-off ground objects.

The outer box allowed the JCATS operator situational awareness of JTLS objects which were either approaching or had potential to effect his operations while significantly reducing the computational overhead for JCATS.

MRM Air Combat Example

A typical MRM experiment would be the integration of the Air Warfare Simulation (AWSIM) and the Modular Semi-Automated Forces (ModSAF) system. AWSIM controls the aggregate flights of aircraft through mission ingress and egress. However, when the missions enter enemy airspace and the unit disaggregated ModSAF controls the attack on the bridge. This allows the application of high-fidelity air defense and target attrition while the aggregate simulation handles the long and tedious ingress and egress legs of the mission.

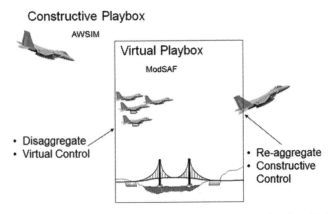

Figure 12-8. MRM Air Combat Example: attacking a bridge with a four-bird sortie

MRM Issues

Consistent History

When disaggregating a unit, the purposes is primarily for some form of engagement or combat. At the virtual level the vehicles are damaged or destroyed on an individual basis. When the virtual engagement is completed the unit may reaggregate to perform other missions at the aggregate level. The damage and attrition experienced at the virtual level may also be aggregated, distributing it throughout the entire aggregate unit.

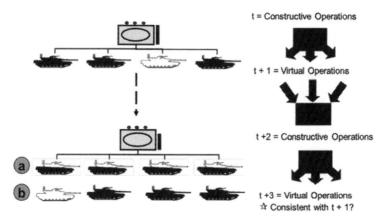

Figure 12-9. Does the aggregate-level show combat vehicle partial damage? Used with permission from the author, Dr. Roger Smith, Copyright 1996-2003.

This does not become an issue until the unit disaggregates a second time. Then the attrition may not be reinstantiated in the virtual world in the same form in which it was originally applied. In this case the destruction if the second vehicle of a platoon may be redistributed in two ways:

1. A different vehicle may be destroyed. Because of computer programming practices this tends to be either the first or last vehicle listed in the software.

2. The damage may be apportioned equally over the entire platoon. This results in four living tanks, each at 75% of their original strength.

Some form of history maintenance is required in MRM systems to prevent this.

Movement

Aggregate and virtual level models encounter different terrain representations, resulting in position calculations that may not be consistent. An aggregate unit may plot a course across terrain that is totally impossible in the virtual world. Or the virtual vehicles may experience a delay due to funneling effect of the terrain features.

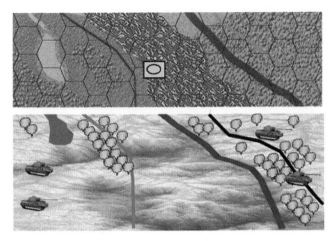

Figure 12-10. Where are the combat vehicles actually located at the aggregate level?

In this situation the aggregate unit will be able to maneuver more freely in all directions and at speeds that cannot be matched by the virtual vehicles. This becomes an important limitation is cases where a unit is partially aggregate and partially virtual, each half trying to remain "attached" to the other.

Light-of-Sight

Line-of-sight is an essential capability for sensor detection and direst fire engagements. Aggregate terrain may lower peaks or minimize the influence of built-up areas in the environmental data, allowing LOS between units that cannot be seen in the virtual world.

Figure 12-11. Does every combat vehicle at the disaggregated level have line-of-sight?

This situation may arise when aggregate movement results in the detection of an enemy unit. The software decides to conduct the engagement at the virtual level and disaggregates the vehicles to perform the combat. However, after the disaggregation has occurred, the two forces cannot see each other and do not know how to conduct the mission they were assigned.

Spreading Disaggregation

In some scenarios a single virtual vehicle can result in the disaggregation of an entire battle front.

A virtual vehicle may interact with an aggregate unit, requesting that it disaggregate for the operation. That unit interacts with other constructive units and makes the same disaggregation request of them. This can cause a domino effect all across the front in which all aggregate units in combat transition to the virtual level based on a single interaction with a single "real" virtual vehicle hundreds of meters away.

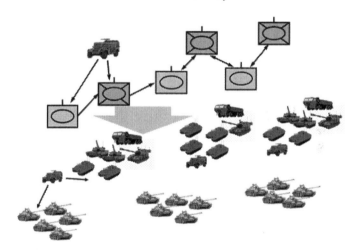

Figure 12-12. The HMMWV leaves the battalion, causing the entire unit to disaggregate. Used with permission from the author, Dr. Roger Smith, Copyright 1996-2003.

Direct Fire Solutions

Several firewalls have been designed to limit the effects of the spreading and cyclic disaggregation just shown. The first firewall calls for only the disaggregation of units that are in direct contact with the virtual vehicle.

Disaggregated vehicles are not allowed to trigger the disaggregation of adjacent units.

Firewall: Direct Contact

You can see that the aggregate unit along the boundary line is in a situation where it must totally break-off engagements with the other aggregate unit in order to transfer itself to the virtual world.

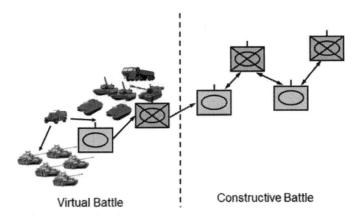

Virtual Battle | Constructive Battle

Figure 12-13. A mechanized company comes into contact with a Red tank company and must entirely disaggregate in the virtual battlefield to fight. The unengaged portion of the battalion remains aggregated in the constructive simulation. Used with permission from the author, Dr. Roger Smith, Copyright 1996-2003.

Firewall: Partial Directed

The second firewall addresses the issue of pushing too much firepower into the virtual world in response to contact with a virtual vehicle. It is possible to determine the size of the virtual threat and disaggregate only enough force to react to it. The remainder of the unit stays in the aggregate world to continue the aggregate level combat.

This approach essentially attrits the aggregate unit and sends that amount of force into the virtual world for combat. Later reaggregation adds the virtual firepower back into the aggregate unit.

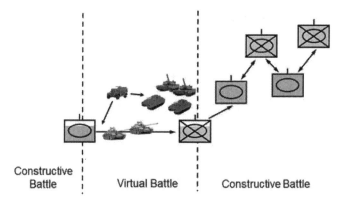

Figure 12-14. A combined arms platoon disaggregates from a mechanized company and fights on the virtual battlefield, without causing further disaggregation from the constructive simulation. Used with permission from the author, Dr. Roger Smith, Copyright 1996-2003.

Firewall: Horizon of Interest

Another form of limited disaggregation is to determine the horizon of interest of the virtual vehicle. Only aggregate units within this area are disaggregated. This allows the transfer of forces to be based on more than just the current contact between units, but prepares forces which may be needed in the battle on the near future.

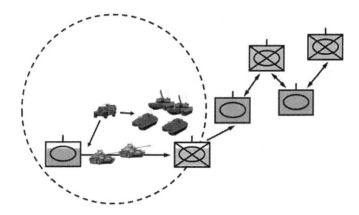

Figure 12-15. This situation is similar to Figure 12-14, except the linear boundary is now a circle with radius equal to the vehicles zone of control. Used with permission from the author, Dr. Roger Smith, Copyright 1996-2003.

Units straddling the boundary of the horizon may apply the same "partial Directed" algorithm described on the previous slide. The horizon of interest may be a weapon range, sensor range, or a measure of the ability of the virtual vehicle to respond to a certain number of enemy vehicles.

Aggregate Location

When vehicles are aggregated into a single unit there are many options for selecting the location of the aggregate unit. It may be placed at the calculated center of mass of the vehicles (the Mean location); or at the middle of the total occupied area (the Median location); or within a grid that contains the largest number of vehicles (the Mode location); or at the location of a key vehicle in the group (such as the commander's vehicle).

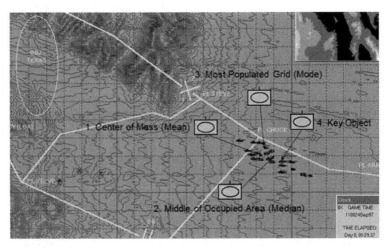

Figure 12-16. Aggregate unit locations may be calculated and depicted differently in different simulations.

Each of these techniques presents an opportunity for inconsistency and for players to take advantage of the algorithm. We will describe two of these here:

1. **Mean Location**. It is possible to construct disaggregation templates that throw many vehicles toward the front of the formation. Upon reaggregation, this template will pull all of the vehicles into an aggregate location that is forward of the original location without moving the forces at all. Repeated aggregation-disaggregation will move the unit across the terrain without burning fuel or being delayed by terrain.

2. **Key Vehicle.** This method is one of the most practical. However, it is open to player "cheating." In the virtual world, the player must only move the command vehicle forward in order to use the aggregation-disaggregation process to move the entire unit forward.

Measure of Consistency

Figure 12-17 is a traditional way depicting consistency. The figure shows initial and final states in the corners of the rectangular diagram. Model A is a high-resolution model and appears in the forward plane; Model B is a lower- resolution model and appears in the rear plane. We assume that the initial aggregate states (2 and 5) are the same; that is, there is a valid way to aggregate from the initial high-resolution state. In this depiction, it is usually said that A and B are "weakly consistent" if the final aggregate states 4 and 6 are the same. That is, if one gets the same result by starting with the high-resolution initial state, simulating with Model A, and then aggregating (1 to 3 to 4), or by starting with the initial aggregate state and simulating with Model B (5 to 6). "Strong consistency" is then usually said to exist if the final high-resolution states are equal (3 and 7), that is, if one gets the same results with paths 1 to 3, or with 5 to 6 to 7. The diagram highlights the fact that information is lost in aggregation, thereby leaving the impression that strong consistency is impossible or quite unusual.

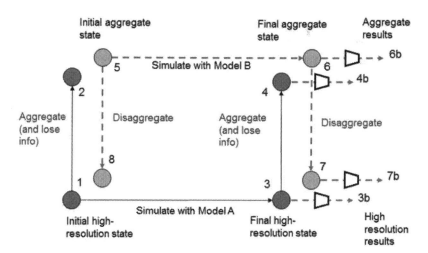

Figure 12-17. *Extracted from Experiments in Multi-Resolution Modeling (MRM), by Paul K. Davis and James H. Bigelow, RAND, 1998 (Davis & Bigelow, 1998).

Paul Davis has improved the diagram so that it includes funnel symbols denoting "projection operators" to emphasize that what matters in the context of an application is some processed version of some portion of the final state, not the full state itself. For example, what matters may be the validity of a graph on a briefing chart, an object's position on a video monitor, or the rank order of some option in terms of cost effectiveness. In each of these cases the "result" being focused upon involves much less information than the final state of a simulation. The figure depicts this in the top-right corner by indicating a final aggregate simulation state with Model B (6), which is then processed by a projection operator to create "aggregate results" (6b). Alternatively, one might have used the higher-resolution Model A to end up with its version of the final aggregate state (4), and then processed that result to achieve result 4b. The question for "weak consistency" is whether 4b and 6b agree adequately. Do the corresponding graphs tell the same story? Do the video-monitor pictures look pretty much the same? And so on. The standards might be lower in program analysis, for example, than in mission rehearsal. The projection operators would also be very different (e.g., the former might involve a weighted ensemble average over future war scenarios, while the latter would be much more focused).

Multiple Levels of Resolution

The definition of Multi-Level Entities puts forward the idea of creating simulation entities that maintain both aggregate and virtual level information about themselves. They are internally responsible for keeping this information consistent. External simulations then request information about the unit at the level of interested.

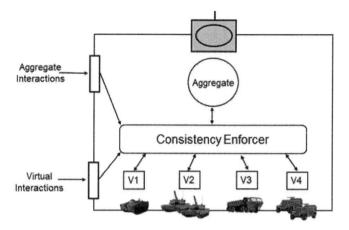

Figure 12-18. Both the aggregate and disaggregate simulations have the same information.

The advantage of this method is that it insures consistency between the aggregate and virtual representations. The disadvantage is that it imposes additional computations for every event that effects a characteristic held in both representations. This method is not useful where broadcast interoperability is used, but may work very well when interoperability is via a publish/subscribe mechanism. The method was first conceived at the University of Virginia and was being implemented in a slightly modified form for the US Army's Warfighter Simulation 2000 system.

Disaggregation Layers

Doctrinal Template. The first layer represents the textbook solution for equipment deployment. This is an instantiation of the placement of subordinate units and equipment according to standard military doctrine. It reflects the type, size, and activity of the unit in a manner very similar to the branch-and-node templates described in the previous section. This is the static form from which adjustments are made in the lower layers.

Historic Events. This layer adjusts the content and status of a unit based on past events. These may include attrition, splitting, merging, and task organization operations. These are the fundamental changes to the unit that are based on events rather than external environment. Each unit may have missing or excess equipment and does not perfectly fit the doctrinal template. Basic mechanistic decisions for handling this situation exist in this layer.

Environment. The effects of weather, night, twilight, and chemical environment are included here. Each of these may result in tighter unit formations and movement precautions that are not experienced during good weather, daylight operations.

Large Feature. Following the placement of objects according to doctrine, history, and environment, the object locations are adjusted according to the presence of large environmental features such as oceans, lakes, swamps, rivers, wooded areas, and cities. This layer does not identify the presence of individual trees and buildings within the features, but merely responds to the border, grade, and altitude of the area covered by the large feature. In some cases, the result is a complete avoidance of the area, as with an ocean or lake; in others it results in the adaptation of the formation to operate in cities or wooded areas. The method may be used for naval forces, in which case the presence of land causes avoidance to keep vessels from beaching or foundering on reefs and shoals. For aircraft, the areas to avoid include the incursion of terrain into the sky and the establishment of flight corridors over certain areas. Air Force disaggregation will keep aircraft out of no-fly zones to avoid destruction and the provocation of enemy forces by incursion into their air space.

Small Feature. This layer takes the previous one to more detail. It accounts for the actual contour of the land and the depths and banks of rivers. This is where adjustments are made to place vehicles on roads and to respond to dynamic terrain features such as bomb craters, and tank ditches. Individual trees, buildings, water towers, and power lines are considered here.

Internal Operations. Following all of the above adjustments, consideration must be given to the operational status and environment of the unit. Its total strength will result in deployment modifications to compensate for attrition on one flank. Vehicles will be moved to account for damage and the depletion of supplies. This includes modifications for the presence or loss of night vision equipment, chemical gear, and other small items used to respond to the environment.

Vehicles without fuel, ammunition, or repair parts may be required to fall to the rear or remain behind while the main body advances. In this level we consider the possibility that the unit will not remain cohesive. It is allowed to break into smaller pieces where each has its own objective and limitations.

External Operations. The presence of enemy forces can cause adjustments in vehicle placement and orientation. This will provide a strong defense, prepare for an eminent engagement, or present a pitched battle. Consideration is also given to the presence and density of non-combatants in the area and the placement of known minefields.

Controller. Finally, the placement of objects may be manually adjusted by orders from simulation controllers or the training audience to produce custom designed patterns. External simulations that are providing specific direction to individual vehicles may also provide these adjustments. This will support the combination of automated units and vehicles that are controlled by other manned simulators.

Battlefield Metaphysics

The legacy view of aggregation-disaggregation has focused on the physical battlefield. It has supported vehicle location, attrition, and movement. However, a modern view of the battlefield should represent assets as information. Each vehicle is actually a container of information about some physical asset. The state of that information and its relationship with other information is the most important characteristic of simulated entities. Simulation models will evolve from "combat heavy" to "information heavy" models. Communication, perception, influence, attitude, and analysis will have more impact on the effectiveness of combat than will the damage levels of tanks.

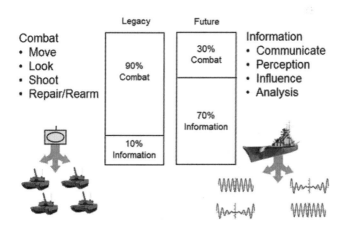

Figure 12-19. Legacy versus future simulations view of aggregation. Used with permission from the author, Dr. Roger Smith, Copyright 1996-2003.

"Steel on Target" becomes "Information on Target." Steel can easily be represented as information, but it is difficult to represent information with steel.

Homeland Security & Humanitarian Operation

Many of the assets and actions within the emergency management domain have similarities with military assets and actions. This may allow the adaptation of military tools to represent this domain. There have been several attempts to demonstrate this type of adaptation. The best known of these is the Plowshares project in which the Janus combat model was modified to represent hurricanes, tornadoes, fires, and the emergency response assets that would handle such events. A more recent project entitled EPICS has attempted a similar transformation of the JANUS simulation. Companies such as Raytheon have also converted some of their internal wargaming tools into emergency management tools and PC combat games like Real War by OCI are being modified to include cooperation between different organizations.

The Civil Emergency Reaction and Responder Training System (CERRTS) was derived from a military simulation and now represents emergency response organizations and disasters that might occur in a city. Several organizations are experimenting with conversions of leaderships training systems like the Joint Combat Analysis and Tactical Simulation (JCATS), JANUS, and Joint Semi-Automated Forces (Joint SAF) which may be able to present command situations to emergency managers as effectively as they have for military commanders.

Commercial-off-the-Shelf (COTS) Simulations

TacOps 4

TacOps 4 is the commercial version of "TacOpsCav 4", an officially issued standard training device of the US Army. It is a simulation of contemporary and near-future tactical, ground, combat between United States (Army and Marine), Canadian, New Zealand/Australian and German forces versus various opposing forces (OPFOR), simulating the Former Soviet Union, China, North Korea etc. Various civilian units and paramilitary forces are also included.

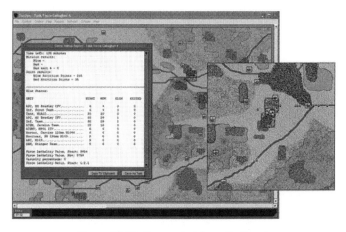

Figure 12-20. Screenshot from TacOps 4

For direct fire weapons, the computer to assess hits and damage usually uses the following guidelines. A basic to-hit probability is calculated based on range to target. The to-hit probability is usually increased if the firer has previously shot at the same target. The to-hit probability is usually decreased if the target is in defilade or entrenched mode, is in smoke, rough, woods, or town terrain, or is moving. The to-hit probability is usually decreased if the firer is moving, suppressed or if the firing weapon is an anti-armor weapon and the target is infantry. The to-hit probability may also be randomly adjusted up or down due to inexplicable factors. The computer considers the final probability and assesses a hit or a miss with a figurative die roll.

If the target is hit, a damage probability is calculated based on a comparison of ammunition effectiveness versus target armor protection. The angle of fire on the target is also considered since armored units normally have the most armor in the front, less on the sides, and least in back. Infantry units also often end up in a makeshift position with most cover to the front, less to the sides, and least in the rear. The computer compares a random number to the damage probability and assesses damage, no damage, or a kill.

Brigade Combat Team

Brigade Combat Team (CT) *Commander* is a real time strategy game that examines the complexities of fighting on the modern battlefield. Developed by Patrick Proctor and his team at ProSim, *BCT Commander* is real enough to be in use as a senior commander training tool for armies around the world.

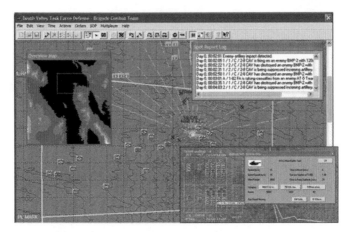

Figure 12-21. Screenshot from Brigade Combat Team

If you've ever bemoaned the lack of realism found in most wargames here's your chance to step up to the plate and see how well you would do in command. BCT Commander is more than a wargame though, it's a complete simulation of land warfare.

Decisive Action

Decisive Action is a modern Division and Corps level simulation that depicts combat with maneuver brigades and battalions along with supporting artillery, airstrikes, electronic warfare, engineer, helicopters, and even pysops units. Units are depicted with official NATO symbology, and US Army official map control measures delineate the battlefield. Scenarios include Germany, SouthWest Asia, and the US Army's National Training Center.

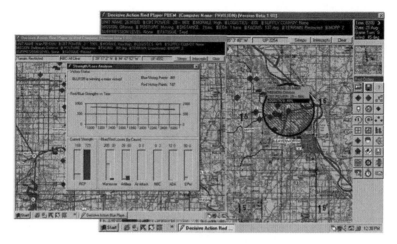

Figure 12-22. Screenshot from Decisive Action

The combat system differs from most conventional wargames. Decisive Action provides each unit with a zone of influence and a Relative Combat Power (RCP). Although, units are not broken down into the number of weapons or vehicles in the organization, they are allocated an RCP based on the unit's size and its main equipment type. For example, a Red Force Tank Brigade equipped with T80B tanks has a RCP of 26, a Red Force Tank Brigade equipped with T72 tanks has a RCP of 20 and a Blue Force M1A2 Tank brigade has a RCP of 34. When two opposing units' zones of influence overlap, combat losses are calculated using the RCP ratio of the units in contact, with adjustments made for their current mission status, the current level of combat effectiveness, the level of logistic support, unit facing, level of morale and fatigue. Losses are reflected in a percentage loss of combat effectiveness.

This page intentionally left blank

References

Works Cited

An Update From The Development Team. (2007, December 7). *America's Army 3.0*.

Recent News From The Development Team. (2007, July 19). *America's Army 3.0*.

Virtual Iraq. (2009). *Virtually Better*.

Adair, B. (2005, February 20). Did the Army get out-gamed? *The St. Petersburg Times*.

Alberts, D. S., Garstka, J. J., & Stein, F. P. (1999). *Network Centric Warfare: Developing and Leveraging Information Superiority* (Second ed.). Library of Congress.

Alberts, D. S., Garstka, J. J., Hayes, R. E., & Signori, D. A. (2001). *Understanding Information Age Warfare*. CCRP Publication Series.

Allen, P. (1992). *Situational Force Scoring: Accounting for Combined Arms Effects in Aggregate Combat Models*. RAND.

Allen, T. B. (1987). *War Games: Inside the Secret World of the Men who Play at Annihilation*. New York: McGraw Hill.

Ambrosiano, N. (2005, October). Largest computational biology simulation mimics life's most essential nanomachine. *Los Alamos National Laboratory News Release*.

Anderson, J. A. (1995). *An Introduction to Neural Networks*. MIT Press.

Arkin, R. C. (1998). *Behavior-Based Robotics (Intelligent Robotics and Autonomous Agents)*. The MIT Press.

Balci, O., Nance, R. E., Derrick, E. J., Page, E. H., & Bishop, J. L. (1990). Model Generation Issues in a Simulation Support Environment. *Proceedings of the 1990 Winter Simulation Conference*, (pp. 257-263). New Orleans, LA.

Barricelli, N. A. (1954). Esempi numerici di processi di evoluzione. *Methodos*, pp. 45-68.

Barricelli, N. A. (1957). Symbiogenetic evolution processes realized by artificial methods. *Methodos*, pp. 143-182.

Barricelli, N. A. (1963). Numerical testing of evolution theories. Part II. Preliminary tests of performance, symbiogenesis and terrestrial life. *Acta Biotheoretica*(16), pp. 99–126.

Birkel, P. A. (1999). SNE Conceptual Reference Model. *Proceedings of the 199 Fall Simulation Conference*.

Boissonnière, F. L. (1999). *An Approach to Design Autonomous Agents within ModSAF*. MSc thesis, RMC, Mathematics and Computer Science.

Bonabeau, E., Dorigo, M., & Theraulaz, G. (1999). *(1999). Swarm intelligence—from natural to artificial systems*. Oxford: Oxford University Press.

Brachman, R. J., & Levesque, H. J. (2004). *Knowledge Representation and Reasoning*. Morgan Kaufmann.

Bracken, J., Kress, M., & Rosenthal, R. E. (1995). *Warfare Modeling*. MORS.

Bray, H. (2003, December 9). New Take On The Game of Life. *Boston Globe*.

Brunett, S., & Gottschalk, T. (1997). *An Architecture for Large ModSAF Simulations Using Scalable Parallel Processors*. California Institute of Technology, Center for Advanced Computing Research, Pasadena, CA.

Caffrey, M. (2000). *History of Wargames: Toward a History Based Doctrine for Wargaming*. StrategyWorld.com.

Caldwell, B., Hartman, J., Parry, S., Al Washburn, & Youngren, M. (2000). *Aggregated Combat Models*. Naval Postgraduate School, Operations Research Department, Monterey, California.

Capitol Reports. (2006, August 22). Gensym G2 Rule Engine Helps CH2M HILL Manage Columbia River Protection Project. *Environment New Link*.

Cares, J. R. (2004). *An Information Combat Model.* Produced for the Director, Net Assessment, Office of the Secretary of Defense.

Clancy, T. (1988). *Red Storm Rising.* New York: HarperCollins.

Crosby, J. L. (1973). *Computer Simulation in Genetics.* London: John Wiley & Sons.

Davis, P. K., & Bigelow, J. H. (1998). *Experiments in Multiresolution Modeling (MRM).* RAND.

Dickie, A. J. (2002). *Modeling Robot Swarms Using Agent-Based.* Thesis, Naval Postgraduate School.

DMSO. (2001). *RTI 1.3-Next Generation Programmer's Guide Version 4.* U.S. Defense Modeling and Simulation Office (DMSO).

Dorigo, M., & Gambardella, L. M. (1997). Learning Approach to the Traveling Salesman Problem. *IEEE Transactions on Evolutionary Computation, 1*(1), p. 214.

Dorigo, M., & Stützle, T. (2004). *Ant Colony Optimization.* MIT Press.

Dupuy, T. N. (1987). *Understanding War: History and Theory of Combat.* Falls Church, VA: Nova.

Elmaghraby, S. E. (1968, June). The role of modeling in IE design. *Industrial Engineering, 6*, pp. 292-305.

Epstein, J. M. (1985). *The Calculus of Conventional War: Dynamic Analysis without Lanchester Theory.* Washington, D.C.: Brookings Institute.

Epstein, J. M., & Axtell, R. (1996). *Growing artificial societies: social science from the bottom up.* Brookings Institution Press.

Fogel, D. B. (1998). *Evolutionary Computation: The Fossil Record.* New York: IEEE Press.

Fowler, B. W. (1996). *De Physica Beli: An Introduction to Lanchestrial Attrition Mechanics, Part III.* Rept. SOAR 96-03, IIT Research Institute/DMSTTIAC.

Fraser, A. (1957). Simulation of genetic systems by automatic digital computers. I. Introduction. *Aust. J. Biol. Sci., 10*, pp. 484–491.

Fraser, A., & Burnell, D. (1970). *Computer Models in Genetics*. New York: McGraw-Hill.

Ganesh, B. (1998). *Return Fire 2 Review*.

Gozel, R. (2000). *Fitting Firepower Score Models to the Battle of Kursk Data*. NPS.

Graham-Rowe, D. (2005, June). Mission to build a simulated brain begins. *New Scientist*.

Griffin, S. P., Page, E. H., Furness, C. Z., & Fischer, M. C. (1997). Providing Uninterrupted Training to the Joint Training Confederation (JTC) Audience During Transition to the High Level Architecture (HLA). *Proceedings of the 1997 Simulation Technology and Training Conference*. Canberra, Australia: U.S. Army STRICOM.

Guilfoyle, C., & Warne, E. (1994). *Intelligent Agents: New Revolution in Software*. Ovum LTD.

Hall, H. E., Shapiro, N., & Shukia, H. J. (1993). *Overview of RSAC system software : a briefing*. RAND Corporation.

Hartley, D. S., & Helmbold, R. L. (1995). Validating Lanchester's Square Law and Other Attrition Models. In Bracken, Kress, & Rosenthal, *Warfare Modeling* (pp. 467-492). Alexandria, VA: MORS.

Hines, J. G. (1990). The Soviet Correlation of Forces Method. In R. K. Huber, H. J. Linnenkamp, & I. Scholch, *Military Stability—Prerequisites and Analysis Requirements for Conventional Stability in Europe* (pp. 185-199). Baden-Baden, FRG: NOMOS Verlag.

Hines, J. G., Petersen, P. A., & III, N. T. (1983, Fall). Soviet Military Theory from 1945-2000. *The Washington Quarterly, 9*(4), pp. 117-137.

Holland, J. H. (1975). *Adaptation in Natural and Artificial Systems*. MIT Press.

Ilachinski, A. (2000). *Towards a Science of Experimental Complexity*. Center for Naval Analyses.

Ilachinski, A. (2001). *An Artificial-Life Approach to War*. Center for Naval Analyses.

ILS. (2009). *Gensym G2*. ILS Process Management.

Jain, A., Mao, J., & Mohiuddin, K. (1996). Artificial Neural Networks: A Tutorial. *IEEE computer, 29*.

Jain, S., & K. Sandeep. (2004). *Graph Theory and the Evolution of Autocatalytic Networks*.

Jain, S., & Sandeep, K. (2004). *Graph Theory and the Evolution of Autocatalytic Networks*.

Johnson, J. (1958). Analysis of Image Forming Systems. *Proc. of Image Intensifier Symposium*, (pp. 249-27).

Kauffman, S. (1995). *At Home in the Universe*. New York: Oxford University Press.

Kiddle, C., Simmonds, R., Williamson, C., & Unger, B. (2003). Hybrid Packet/Fluid Flow Network Simulation. *Workshop on Parallel and Distributed Simulation, Proceedings of the seventeenth workshop on Parallel and distributed simulation*, (p. 143).

Koopman, B. O. (1946). *OEG Report 56 on Search and Screening*. Navy Department, Operations Evaluation Group. Washington, D.C.: Office of the Chief of Naval Operations.

Kuhl, F., Weatherly, R., & Dahmann, J. (1999). *Creating Computer Simulation Systems: An Introduction to the High Level Architecture*. Prentice Hall.

Laird, J. E., Jones, R. M., & Nielsen, P. E. (1998). Lessons learned from TacAir-Soar in STOW-97. *Proceedings of the Seventh Conference on Computer Generated Forces and Behavioral Representation*. Orlando, FL.

LaMothe, A. (1999, July 31). Building Brains Into Your Games. *Gaming*.

LaMothe, A. (1999, October 7). Neural Netware. *Neural Networks*.

Lamport, L. (1978, July). Time, Clocks, and the Ordering of Events in a Distributed System. *Communications of the ACM, 21*(7), pp. 558-565.

Lauren, M. K., & Stephen, R. T. (2002). Map-aware Non-uniform Automata (MANA) a New Zealand Approach to Scenario Modelling. *Journal of Battlefield Technology, 5*(1).

Leeson, B. (2011). Origins of the Kriegsspiel. *Kriegsspiel News*.

Lehman, J. F., Laird, J., & Rosenbloom, P. (2006). A Gentle Introduction to SOAR, an Architecture for Human Cognition: 2006 Update.

Logsdon, J., Nash, D., & Barnes, M. (2008). *One Semi-Automated Forces (OneSAF) Capabilities, architecture, and processes*. TRADOC Analysis Center.

Matthews, J. (2000). An Introduction to Neural Networks. *Generation5*.

Matute, E. B. (1970). Birth and Evolution of War Games. *Military Review, 50*(7), p. 53.

McIntosh, G. C., Galligan, D. P., Anderson, M. A., & Lauren, M. K. (2008). Recent Developments in the MANA Agent-based Model. *The Scythe*(1), pp. 38-39.

Nance, R. E. (1971, September). On Time Flow Mechanisms for Discrete Event Simulations. *Management Science, 18*(I), pp. 59-93.

NASA. (2009). *The Shuttle Radar Topography Mission*. California Institue of Technology, Jet Propulsion Laboratory. National Aeronautics and Space Administration.

Nicolas, M., Frédéric, G., & Patrick, S. (2010). *Artificial Ants*. Wiley-ISTE.

Nikolakopoulos, K. G., Kamaratakis, E. K., & Chrysoulakis, N. (2006, November 10). SRTM vs ASTER elevation products. Comparison for two regions in Crete, Greece. *International Journal of Remote Sensing, 27*(21).

Osinga, F. (2006). *Science, Strategy and War: The Strategic Theory of John Boyd*. Routledge.

Parry, S. H., & Hatman, J. K. (1992). *Airland Combat Models II: Aggregated Combat Modeling.*

Pawlowski, T. (2005). Applying Multi-Agency Executable Architectures to Analyze a Coastal Security Operation. *10th International Command and Control Research and Tachnology Symposium.*

Perry, D. C. (2004, May 31). Full Spectrum Warrior Review. *IGN.*

Pimental, K., & Blau, B. (1994). Teaching Your System To Share. *IEEE computer graphics and applications, 14*(1), p. 60.

Pollack, M. E. (2006). *Intelligent Assistive Technology: The Present and the Future.* Computer Science Distinguished Lecture , UMass Amherst Campus, Computer Science , Amherst.

Prochnow, D. L. (2002). JTLS-JCATS: Design of a Multi-Resolution Federation for Multi-Level Training. *Procedings of the 2002 Fall Simulation Interoperability Workshop.* Orlando, FL.

Proctor, M. D., & Gerber, W. J. (2004, April). Line-of-Sight Attributes for a Generalized Application Program Interface. *JDMS, 1*(1), 43–57.

Rechenberg, I. (1973). *Evolutionsstrategie.* Stuttgart: Holzmann-Froboog.

Rheingold, H. (1992). *Virtual reality.* New York, N.Y.: Simon & Schuster.

Robinett, W. (1994). Interactivity and Individual Viewpoint in Shared Virtual Worlds: The Big Screen vs. Networked Personal Displays. *Computer Graphics, 28*(2), p. 127.

Rosenbloom, P. S., Laird, J. E., & Newell, A. (. (1993). *The Soar Papers: Research on.* Cambridge, MA: MIT Press.

Russell, S. J., & Norvig, P. (1995). *Artificial Intelligence: A Modern Approach* (1st ed.). Prentice Hall.

Schwefel, H.-P. (1977). *Numerische Optimierung von Computor-Modellen mittels der Evolutionsstrategie : mit einer vergleichenden Einführung in die Hill-Climbing- und Zufallsstrategie.* Stuttgart: Birkhäuser.

Schwefel, H.-P. (1981). *Numerical optimization of computer models.* New York: Wiley.

Seaman, J. (2001). *Artificial Neural Networks Vs Biological Neural Networks, Can A Computer Learn?* University Of Central Lancashire.

Seidel, D. (1993). *Aggregate Level Simulation Protocol (ALSP) Program Status and History.* McLean, VA: The MITRE Corporation.

Smith, L. (1996). An Introduction To Neural Networks. University Of Stirling.

Smith, R. (1998). Simulation Article. *Modelbenders.com.*

Smith, R. (2006). *Military Simulation Techniques & Technology.* Model Benders, LLC.

Sokolowski, J. A., & Banks, C. M. (2009). *Principles of Modeling and Simulation.* Hoboken, NJ: Wiley.

Strickland, J. S. (2004). Introduction. In *Combat Modeling* (p. 1.13). US Army Logistics Management College.

Strickland, J. S. (2004). *ORSA Processes.* Fort Lee, VA: US Army Logistics Management College.

Strickland, J. S. (2005). Introduction. In J. S. Strickland, *Combat Modeling.* Fort Lee, VA: Army Logistics Management College.

Strickland, J. S. (2005). Multiresolution Modeling. In J. S. Stricklnd, *Combat Modeling.* Fort Lee, VA: US Army Logistics Management College.

Strickland, J. S. (2006). An Agent-Based Simulation of Satellite Utility For the US Army Future Force. In S. T. S. Robinson (Ed.), *Proceedings of the 2006 OR Society Simulation Workshop.*

Strickland, J. S. (2006). *Networked Virtual Simulation.* Lesson notes for Modeling and Simulation Professional Program, University of Alabama in Huntsville, Center for Modeling, Simulation, and Analysis.

Strickland, J. S. (2009). Mathematical and Heuristic Models of Combat. *Interservice/Industry Training, Simulation and Education Conference (I/ITSEC)*. Orlando, FL.

Strickland, J. S. (2010). *Discrete Event Simulation Using ExtendSim 8*. Lulu, Inc.

Strickland, J. S. (2010). *Missile Flight Simulation: Surface-to-Air Missiles* (First ed.). Lulu, Inc.

Strickland, J. S. (2011). *Fundamentals of Combat Modeling*. Lulu, Inc.

Strickland, J. S. (2011). *Mathematical Modeling of Warfare and Combat Phenomenon*. Lulu, Inc.

Strogatz, S. (2007). The End of Insight. In *In Brockman, John. What is your dangerous idea?* HarperCollins.

Taylor, J. G. (1983). *Lanchester Models of Warfare*. Arlington, VA: Operations Research Society of America.

Taylor, J. G., & Parry, S. H. (1975, May-June). Force-Ratio Considerations for Some Lanchester-Type Models of Warfare. *Operations Research, 23*, pp. 522-533.

Taylor, J. G., Yildirim, U. Z., & Murphy, W. S. (2000). Hierarchy-of-Models Approach for Aggregated-Force Attrition," , , . In J. A. Joines, R. Barton, K. Kang, & P. A. Fishwick (Ed.), *Proceedings of the 2000 Winter Simulation Conference* (pp. 925-932). Orlando, FL: Winter Simulation Conference.

Taylor, J. R. (1999). *An Introduction to Error Analysis: The Study of Uncertainties in Physical Measurements*. University Science Books.

Tchoukanski, I. (2005). *Triangulated Irregular Networks*.

Testa, B. M. (2008, May 26). America's Army' provides an enterprise platform for Army training. *Defense Systems*.

Turse, N. (2003, October 17). Bringing the War Home: The New Military-Industrial-Entertainment Complex at War and Play. *CommonDreams.org*.

Volimerhausen, R. H., Jacobs, E., & Driggers, R. G. (2003). New metric for predicting target acquisition performance. In G. C. Holst (Ed.), *Proceedings of SPIE: Infrared Imaging Systems: Design, Analysis, Modeling, and Testing XIV, 5076*, pp. 28-40.

Waldner, J.-B. (2007). *Inventer l'Ordinateur du XXIème Siècle.* London: Hermes Science.

Waldner, J.-B. (2008). *Nanocomputers and Swarm Intelligence.* London: ISTE John Wiley & Sons.

Watson, J. G. (1997). *Researchers stage largest Military Simulation ever.* Caltech, Jet Propulsion Laboratory.

Weatherly, R. M., Wilson, A. L., Canova, B. S., Page, E. H., Zabek, A. A., & Fischer, M. C. (1995). Advanced Distributed Simulation through the Aggregate Level Simulation Protocol. U.S. Army STRICOM.

Wilson, A. (1968). *The Bomb and the Computer.* London: Barrie & Rockliff, Cresset P.

Wilson, A. L., & Weatherly, R. M. (1994). The Aggregate Level Simulation Protocol: An Evolving System. *Proceedings og the 1994 Winter Simulation Conference*, (pp. 781-787). Lake Buena Vista, FL.

Index

#

gi *See* probability of detection

3

3-DOF 98

A

A* Search algorithm 93
accreditation 60
acquisition 2, 4, 20, 29, 59, 66, 105, 107, 108, 109, 118, 121, 124, 126, 127, 151
adjacency matrix... 135, 138, 139, 140, 144, 145, 146
Advanced Concept Requirements 4, 14, 15
agent 192, 193
agent architecture 29, 195, 196
Agent-based simulation 12
agents 33
aggregate .... ii, 44, 46, 113, 165, 188, 235, 237, 243, 244, 245, 247, 248, 249, 250, 251, 252, 253, 254, 255, 256
Aggregate Level Simulation Protocol .. 13, 35, 38, 42, 43, 44, 45, 46, 47, 48, 49, 51
aggregate-level simulation 235
aggregate-level simulations ... 246
aggregation xii, 16, 165, 181, 235, 238, 243, 244, 246, 253, 254, 258
aiming point 159
Air Warfare Simulation 42, 44, 247
ALSP Broadcast Emulator. 43, 48, 51

ALSP Common Module 43, 48, 49, 50, 51
America's Army 154, 216, 217, 218, 232
analytical simulation 16
ant colony optimization 224
Army Capabilities Analysis 177
Army Operations Research Office 16, 17
Artificial Ants 226
artificial intelligence 190, 191, 194, 197, 199, 200, 201, 202, 205, 206, 207, 208, 221
Artificial Neural Networks 210
Asymmetrical Warfare Virtual Training Technology 232
ATLAS theater level simulation 175, 176, 180
Attrition Calibration model .. 173, 177, 239, 240, 241, 242, 243
attrition process 152, 165
attrition-rate coefficient 166
autocatalytic sets 139
automatic route planning 91
avatar 99, 100, 218

B

bald earth movement 89
Battle model 235
BC-2010 232
BCT Commander 260
behavioral agents 30, 32
behavioral modeling 66, 190, 191
behavioral models 30
Bresenham LOS algorithm 76
Brigade Combat Team 260
broadcast messages 132
bSerene 207, 208

C

Campaign model 235
CARMONETTE 17
CASTFOREM 59, 76, 82, 105, 125, 166
CENTCOM 214
CEP ....*See* circular error probable
CERRTS 259
characteristic equation 169
Chess 6
circular error probable 160, 161, 162
classification 108
CLIPS 213, 214
Close Combat Tactical Trainer 23, 28, 67, 77, 91
Coefficient of Networked Effects 140
combat attrition 3, 87, 165
combat simulations 46, 78, 111, 114, 124
COMBAT XXI 59
Command and Control Simulation Interface Language 35
communications modeling 129
communications networks 130, 131
Compact Terrain Database 74, 75
Computer aided exercise interacted tactical communications simulation 150
computer model 7, 227
computer network 24, 36, 147
computer simulation . 7, 8, 17, 51, 205, 217
computer-generated force 19, 24, 25, 27, 190

Concepts Evaluation Model 18, 59, 188, 239, 240
constructive simulation. 15, 19, 26, 27, 35, 79, 86, 151, 251, 252
continuous dynamic simulation 11
continuous looking model 119
Corps Battle Simulation 17, 18, 42, 44, 59, 79
Correlation of Forces Method 176, 177, 178
correlation of forces model 176
COSAGE 239, 240
credibility 61
cross-range error probable 160

D

damage points 153, 154
DARPA .. 19, 21, 22, 25, 26, 36, 37, 44, 45, 46
Data management 45, 46
data processing 70, 93, 194
deciders 133, 134, 135
Decisive Action 261
deliberative agent 194, 196
Desert Storm 18, 22, 41
detection 104, 106, 107, 109, 122, 124
Detection Map 105, 126
detection range 102, 107
detection rate function 124
detection rates 104
deterministic 10, 18, 59, 112, 173, 200
differential equation 165, 167, 168
differential equations 11, 119, 238
direct message passing 131

disaggregation. xii, 244, 250, 252, 253, 254, 257, 258
discrete event simulation 11, 49, 231
Distributed Interactive Simulation 13, 23, 24, 25, 29, 35, 36, 37, 38, 39, 40, 43, 51, 83, 234
Doctrinal Template 256
dynamic 10, 25, 26, 48, 67, 70, 81, 82, 137, 139, 144, 213, 231, 257
dynamic environments 81
dynamic simulation 10, 12
dynamic terrain 81
DYNTACS 76, 78, 105, 125

E

Eagle 59, 93
eigen-solution 238
eigenvalue 138, 169, 170, 180, 181, 183, 184, 187
eigenvalue method 181
eigenvector 169, 170
EINSTein 228, 229, 230, 231
elevation posts 74, 75, 76, 77, 125, 245
engagement model 235
engagement modeling 151
engineering models 235
engineering simulation 16
entity-level simulations 246
Entropy-Based Warfare is 233
environmental modeling 66, 67
EPICS 259
error 158
error probable 158, 160
evolutionary algorithms 205
Evolving Neural Networks 210
expert system 212, 213

explicit LOS model 115
exponential distribution 125
Extended Air Defense Simulation
.... 18, 59
Extensive aggregation 238

F

Federation Object Model .... *See* FOM
Fields of Battle 204, 205
Finite State Automation 198
finite state machines 190, 198, 199, 200, 206
Fire Power Index 186
firepower score 173, 174
firepower score attrition models
.... 175
fitness function 205
FOM 54, 56
force ratio 165, 173, 174, 180, 185, 186
force ratio attrition models ... 165, 173
Formal Aggregation 237
Full Spectrum Warrior 219, 221
funneling effect 248
fuzzy logic 190, 201, 202
fuzzy machines 203
fuzzy rules 203

G

G2 Expert Systems Development Environment 213
genetic algorithm .. 205, 206, 208, 209, 230, 231
ghosts 48, 99, 100
giant component 142, 143
glimpse interval 120
glimpse model 119, 124, 125
goal-based agent model 196

grid hopping 86
grid registration 97
grid terrain 75, 77

H

Half-Life 200
health points 153
heuristic combat models 3
heuristics .... ii, 3, 191
hexagon 73, 87, 88, 101, 115, 116
hexagons 78, 79, 88, 89, 116, 245
hidden nodes 147
High Level Architecture 13, 29, 35, 38, 45, 51, 52, 53, 54, 55, 56, 57, 83, 234
high-resolution models 239, 243
human factors .... xii, 22

I

identification 104, 108
infantry-fighting vehicle 19, 25, 163
influencers 133, 134, 135, 136, 145, 146
Information Age combat model 133, 138, 144
Information Age Warfare 63
initial conditions 8, 172
Institute for Defense Analysis Gaming Model 185, 187
intelligent agents.... 190, 192, 197
Intensive aggregation 238
intermittent LOS 118
intermittent search 122
interpolation 76, 77, 125
intervisibilty range 117
Irreducible Semi-Autonomous Adaptive Combat 211, 228, 229

ISAAC .... *See* Irreducible Semi-Autonomous Adaptive Combat

J

Janus 76, 77, 78, 82, 151, 162, 259
JCATS .... *See* Joint Conflicts And Tactical Simulation
JMACE ...... *See* Joint Military Art of Command Environment
Joint Conflicts And Tactical Simulation.... 90, 105, 126, 234, 245, 246, 259
Joint Military Art of Command Environment 214
Joint Simulation System 57
Joint Tactical Level Simulation 246
Joint Theater Level Simulation 79, 150, 246
Joint Training Confederation ... 41, 42, 43, 57, 58
Joint Warfare System 59, 113, 114, 201
Joshua Epstein 172, 188, 227
JSIMS .... 59, *See* Joint Simulation System
JTLS .... *See* Joint Theater Level Simulation
JWARS ..*See* Joint Warfare System

K

kill categories 162, 163
kill probability .*See* probability of kill
kill rate 166
kill rates 182
killer-victim scoreboard 182, 242
Kriegspiel 1, 57

L

Lanchester equations 133, 166, 175, 188
Lanchester-type combat model 167
LEP .... *See* linear error probable
limited disaggregation 252
linear error probable .. 158, 160
Line-of-sight 76, 77, 98, 107, 109, 110, 111, 112, 113, 114, 115, 116, 117, 118, 119, 125, 249
Line-of-sight algorithm 114
Line-of-sight blockage 114
links 133, 134, 140, 143, 144
live simulation 16, 27
low-resolution 166, 239, 243

M

M1 Abrams 19, 25
M2 Bradley 19, 25
MÄK Technologies 215, 232
man packed radios 150
MANA .... *See* Map Aware Non-uniform Automata
Map Aware Non-uniform Automata 210, 211, 212
Markov chain model 118, 190, 200, 201
Markov P_K table 155
MARSS .... *See* Multi-Agent Robot Swarm Simulation
mathematical model .... ii, 8, 119, 200
mathematics ...... ii, 3, 10, 124, 125, 132, 133, 138, 199, 200
MATLAB® and Simulink™ 12
mean 157, 158
mean time to detection 125
measure of adaptability 144

measures of effectiveness 124, 125
medium-resolution models .. 166, 239, 240
military exercises 1, 6
military modeling xii
military models xi
Military Operations Research Society 18
military simulation .. xi, xii, 1, 2, 6, 85, 191, 202
military simulation environment 2
Mission model 235
mobility kill 151, 163
model 4, 5
modeling and simulation ..... xii, 3, 38, 56, 99
ModSAF 19, 20, 24, 25, 26, 27, 59, 74, 77, 209, 247
MOE .... *See* measures of effectiveness
Monopoly 4
Monte Carlo simulation . 163, 164
movement-points 87
MOVES institute 232
Multi-Agent Robot Swarm Simulation 231
Multi-Level Entities 255
multi-resolution modeling 66, 234, 243, 245, 246, 247, 248, 254
Multi-Resolution Multi-Perspective Modeling 234

N

National Simulation Center 214
navigational errors 158
Network Centric Warfare . 62, 92, 137

network evolution 141
network simulators 147
neural network 147, 203, 204, 205, 210
Neutrality Rating 144
Night Vision Electro-Optical Lab 105, 126, 127
nodes 133, 134, 140, 141, 143, 144
normal curve 158
normal distribution 159
NVEOL.. *See* Night Vision Electro-Optical Lab

O

O(N) problem 98
Object Model Template 51, 52, 53, 54, 55, 56
OMT.. *See* Object Model Template
OneSAF 20, 27, 28, 29, 30, 31, 33, 67, 74, 77
OneSAF Testbed Baseline... 20, 77
Operation Iraqi Freedom 214
optimization 1, 192, 205, 206, 226

P

parameters xi, 8, 19, 26, 113, 160, 161, 209, 240, 241
PDU 38, 39
perfect communications 129, 130
perfect detection 101
performance. 1, 5, 7, 8, 16, 27, 60, 70, 71, 97, 98, 141, 147, 150, 157, 218, 231
performance characteristics 16
peripheral nodes 144
Perron-Frebonius eigenvalue 138, 139, 140
pheromones 224, 227

physical agents 30, 32
physical modeling 66
physical models 30
physics-based movement .. 86, 91
piston model 18, 87, 173, 188
PLOS .... *See* probability of line-of-sight
point-fire attrition equation .. 243
polyagent modeling 99
Polygon Attribute Table 75
polygons 23, 74, 75, 77, 92
potential/antipotential method 176, 180, 181
precision engagements 157
probabilities 158, 159
probability density function. 157, 158, 159
probability of destruction 161
probability of detecting . 123, 124
probability of detection .. 34, 102, 103, 120, 121, 122, 124, 125
probability of kill .... xi, 154, 155, 156, 162
probability of line-of-sight 111, 112, 113, 114, 115, 116
probability of line-of-sight curve 112, 113
Proper Aggregation 236, 237
proximity detection 101

Q

Quake 199
Quantified Judgment Model .. 176, 188

R

radar cross section 101, 106
radio access points 150
radio frequency 134
Rainbow Six 215, 216, 217

random numbers .. 122, 127, 155, 157, 163
range error probable 158, 160
reactive agent 194, 196
reaggregation 251, 253
real-time .. 20, 21, 22, 24, 37, 213, 221, 231, 233
recognition 104, 108
relative combat power 262
Research, Development, and Acquisition 4, 14, 15
Resolution 243
Return Fire 2 .... 208, 209
RTI. *See* Run-Time Infrastructure
Run-Time Infrastructure ... 51, 52, 53, 54, 55, 56

S

search space 122, 210
SEDRIS 35, 84
semi-automated force.. 19, 26, 27, 28, 40, 59, 157, 190, 259
shooter .. 134, 151, 153, 157, 165, 208, 216, 217, 221, 232
Shuttle Radar Topography Mission 71, 72, 73
SIMNET 19, 20, 21, 22, 23, 24, 25, 26, 35, 36, 37, 38, 44, 46, 51
simulation 4
Simulation Object Model ...... *See* SOM
simulators ... 1, 4, 9, 11, 19, 20, 21, 22, 24, 25, 27, 28, 35, 36, 37, 39, 43, 45, 51, 52, 86, 258
single-shot probability hit or kill 159
SOAR .... *See* State Operator and Response
Software Engineering Directorate 218

solution vector 185
SOM 54, 55, 56
sparse matrices 137
spatial frequency 126
Spearhead 215
standard deviation 157, 158
State Operator and Response 221, 222, 224
Stigmergy 224, 225
stochastic 2, 59, 107, 112, 119, 121, 127, 200
stochastic models 10
STRICOM 20, 26, 219
sub-network 137
Sugarscape 227, 228
swarm 99, 225, 231
swarm theory 231
synthetic environments 27
Synthetic Natural Environment 69
Synthetic Theater of War .. 26, 81, 223
System Evaluation and Analysis Simulation (SEAS) 13
systematic error 159, 160

T

TacAir-Soar 223
TacOps 4 .... 259, 260
Tactical Air Combat Maneuver 68
Tactical Simulation Model. 42, 44
TACWAR 59, 174, 180, 187
Tank Sore 223
target 159
target acquisition .. 103, 104, 107, 108, 119, 123, 182
target engagement 3
target search 122
terrain elevation points 82

terrain type.......90, 102, 111, 112, 116
testing simulation...................... 16
The Universal Military Simulator
....202
Theater..........59, 79, 81, 150, 246
Theater model...........................235
time management....... 46, 48, 49
TIN......*See* Triangulated Irregular Networks
topology smart............................... 92
TRADOC... 20
Training, Exercise, Military Operations........................ 4, 14, 15
Triangulated Irregular Networks
......................73, 74, 75, 77, 114

U

Unreal Engine..................... 216, 217
Unreal Tournament.....................217
urban terrain............ 110, 111, 114
utility based agent model........196

V

validation...................................27, 60

Vector In Command 59, 119, 166, 188
verification 1, 60
Virtual Cell Layout............149, 150
virtual simulation......................16
vulnerability. 26, 30, 32, 163, 243

W

warfare models................................. 1
wargame.. 4
wargames....ii, 5, 6, 15, 17, 35, 43, 57, 58, 78, 86, 87, 89, 201, 204, 215, 261, 262
wargaming........5, 37, 41, 57, 214, 259
Warrior Preparation Center.....35, 41, 45
WARSIM........................ 59, 188, 201
Weapon Effectiveness Index. 176, 177
Weapon Unit Value 177
weapons system............................... xi
Weighted Unit Value 176

X

Xbox.............................. 217, 220, 221

Made in the USA
Monee, IL
03 November 2019